Waste Biomass to Alternative Fuels

Scrivener Publishing
100 Cummings Center, Suite 541J
Beverly, MA 01915-6106

Publishers at Scrivener
Martin Scrivener (martin@scrivenerpublishing.com)
Phillip Carmical (pcarmical@scrivenerpublishing.com)

Waste Biomass to Alternative Fuels

Prakash Kumar Sarangi

Scrivener Publishing

WILEY

This edition first published 2025 by John Wiley & Sons, Inc., 111 River Street, Hoboken, NJ 07030, USA and Scrivener Publishing LLC, 100 Cummings Center, Suite 541J, Beverly, MA 01915, USA

For more information about Scrivener publications please visit www.scrivenerpublishing.com.

Wiley Global Headquarters
111 River Street, Hoboken, NJ 07030, USA

For details of our global editorial offices, customer services, and more information about Wiley products visit us at www.wiley.com.

The manufacturer's authorized representative according to the EU General Product Safety Regulation is Wiley-VCH GmbH, Boschstr. 12, 69469 Weinheim, Germany, e-mail: Product_Safety@wiley.com.

Library of Congress Cataloging-in-Publication Data

ISBN 9781119602088

Background Cover Image: Generated with AI using Adobe Firefly
Inset images courtesy of Wikimedia Commons
Cover design by Russell Richardson

Set in size of 11pt and Minion Pro by Manila Typesetting Company, Makati, Philippines

Printed in the USA

10 9 8 7 6 5 4 3 2 1

Dedicated
to
My beloved Lord

Contents

Preface

Environmental sustainability is a growing need due to increasing population, leading to dependency on fossil fuels and other such products. Due to the industrial revolution and increase of automobiles forcing too much dependency on fossil-based fuels, alternative sustainable sources with less environmental impact must be sought not only for future energy issues but also to build sustainability. This book is an enthusiastic celebration of supplementing waste biomass, having challenging alternatives to fossil fuel derivatives towards sustainability. This book can inspire researchers to find a new pathway to meet energy demands with the help of novel biotechnological interventions. It provides a solid platform with updated information regarding bioconversion technologies for second-generation biofuel production and illustrates diverse chapters with innumerable information, emphasizing different forms of alternative energy. This book enriches readers' knowledge in understanding biprocessing tools and the processes for the production of sustainable energy, focusing on multidisciplined researchers and students in the field of bioenergy, biotechnology, microbiology, chemical engineering, and fuel processing. This book will provide a concrete platform for socio-economic parameters about bioenergy utilization from waste biomass toward sustainability, including knowledge about the relationship of biofuels with waste biomass.

I hope this book will satisfy most needs of academicians and researchers globally in the field of bioprocess technology and applied biotechnology toward sustainable energy and fuel. I take this opportunity to express my sincere thanks to my family members who have put their enthusiastic efforts bestowing most valuable recent research information in completing the book manuscripts. I express my hearty thanks to Mr. Phil Carmical, Scrivener Publishing, LLC, for his kind and timely help, supplying full logistic support toward the publication of this book.

Dr. Prakash Kumar Sarangi
Scientist,
College of Agriculture
Central Agricultural University
Imphal, Manipur, India

Major highlights of this book

- ➢ The production of biofuels from waste biomass through different generations and their process protocols
- ➢ Recent advancements in the methods of biofuel production in various generations with illustrations
- ➢ The valorization of waste biomass to bioethanol and the advancement processes

- ✓ Updated information about fermentation methods in the production of bioethanol that can be used as an alternate source for biofuel
- ✓ Value addition of wastes from various sources, citing their utilization in biohydrogen, biobutanol, biomethanol, bioethanol, and biohythane
- ✓ Biofuel productions and applications with suitable illustrations

- ❖ Illustrations about the sustainability and the economical perspective on biofuels
- ❖ Biotechnological interventions and algal supplementations into biofuel context
- ❖ A showcase for mutidisciplinary students, faculties, and researchers in the field of microbiology, biochemical engineering, and bioprocess engineering
- ❖ Personnel like environmental engineers, policy makers and social workers may be interested in the contents of the book.

1

Waste Biomass to Alternative Fuels: An Overview through Different Generations of Biofuels

Prakash Kumar Sarangi

College of Agriculture, Central Agricultural University, Imphal, Manipur, India

Abstract

The world mainly depends on fossil fuels as energy sources for automobiles and industrial sectors. The uncertainty in the supply, rising demand for energy, and some adverse environmental impacts of crude fossil fuel force researchers around the world to find alternative sources of energy. Fossil fuels not only create environmental problems but will also diminish in the future. Biobased energy sources are abundant in nature and can be explored to mitigate the problems caused by fossil fuels. It is time to focus on new alternative biobased energy approaches toward research and development as further processing technologies reform a sustainable bioeconomy. Sustainability in energy systems can be the outcome of biofuel and its technologies. Lignocellulosic biomass is seen as a major driving force for renewable energy sources worldwide, serving as a promising renewable source for prospect fuel. Depending on the feedstock and processing technologies, biofuels are categorized into four generations. The first generation of biofuels deal mainly with the production of bioethanol food-based edible biomass, whereas the second generation involve the production of various biofuels from waste biomass having no conflict on food versus fuel. Algal biomass, sewage sludge, and municipal solid wastes are used for the production of biofuel positioning in the third generation. The applications of recombinant technologies for algal biomass toward the production of biofuels also have great potential, being the fourth generation of biofuel. This chapter describes a broad analysis of the four different generations along with their future perspectives.

***Keywords*:** Biomass, biofuel, bioethanol, generations, sustainable, fossil fuel

1.1 Introduction

In most power sectors, fossil-based energy sources are utilized as the primary fuels for power purposes. The utilization of petroleum and other fossil-based liquid fuels around the world was about 86 million barrels per day (BPD) in 2008, and there is a prediction of an increase to 98 million BPD in 2020 and 112 million BPD in 2035 (USEIA, 2011). Problems associated with such fossil fuels are not just environmental issues, considering the instability in price and select availability. Other limitations such as legislative restrictions and fluctuations of the fossil price play vital roles using non-renewable energy sources, which force the exploration of biotechnology methods for biofuels production. Moreover, a sustainability-based economy and development are needed for the eco-friendly environment, which would not produce any pollutants harmful to the environment (Sarangi *et al.*, 2020b). Therefore, new grounds of research must be generated to explore novel renewable sources of energy and value-added chemicals that are economic and environment-friendly (Sarangi *et al.*, 2009a; Sarangi *et al.*, 2009b; Sarangi *et al.*, 2010; Sarangi and Sahoo, 2010a; Sarangi and Sahoo, 2010b; Sarangi and Sahoo, 2010c). Hence, green emerging technologies for conversion of plant-based biomass and other organic waste into various biofuels can be considered for the mitigation of a potential energy crisis. Hence, the substitution of fossil fuels with biofuel is a vibrant alternative that is not only cost effective but also eco-friendly. Various by-products produced by agro-industries including corncobs, pineapple wastes, sugarcane bagasse, bran, and straw have been used as cheap substrates for bioconversion to obtain fuels and value-added biochemicals by an environment-friendly bioprocess technologies (Sarangi *et al.*, 2019; Sarangi and Nanda, 2019a; Sarangi and Nanda, 2019b; Sarangi and Nanda, 2019c). A biobased energy system plays a significant role in mitigating fossil-based fuels on a large scale and shares energy applications for chemical and material production in the future. The main global objective lies in the exploitation of clean, green, sustainable alternative sources for production of fuel, power, heat, and value-added chemicals from renewable waste biomass (Sarangi and Nanda, 2018; Sarangi and Nanda, 2020).

Global interest is driven towards exploring more renewable sources that can be used for producing biofuels. Various forms of biofuels viz., bioethanol, biobutanol, biopropanol, biogas, and biodiesel may be attended forms such as biobased renewable sources towards the mitigation of fossil-based environmental issues (Nanda *et al.*, 2020; Pasin *et al.*, 2020; Siang *et al.*, 2020;

Bhatia *et al.*, 2020; Sarangi *et al.*, 2020). Lignocellulosic biofuel has great potential in becoming transportation fuel and gaining significant research in this area can be emphasized towards the development of alternative and eco-friendly economical technologies (Perlack *et al.*, 2005). Waste biomass can be utilized for the production of fuels and valued-added biochemicals by different kinds of methods, like thermochemical and biochemical technologies. Various biofuels and their generations along with production strategies are summarized in this chapter. Also, the applications of a wide array of waste biomass for alternative fuel production in four generations are also discussed.

1.1.1 Lignocellulosic Waste Biomass

Major sources of lignocellulosic waste biomass are (1) agricultural residues such as corn stover and rice straw; (2) forest residues such as woods, branches, and foliage; (3) energy crops such as switch grass and yellow poplar; and (4) cellulosic waste such as municipal solid waste and food waste (Parisuthan *et al.*, 2014). Agro-industrial residues like peels, husk, straws, bagasse, organic waste, lignocellulosic biomass, municipal waste are utilized for second-generation biofuels. Wastelands could be employed to produce feedstock for these types of fuels. Perennial grasses are also employed for the production of second-generation biofuels. They include miscanthus, switchgrass, and reed canary. Agro-residues comprise wheat straw, corn leaves, cobs, stalks, and sugarcane bagasse, even as forestry residues incorporate logging residues and wood. These feedstock are a rich source of organic material in the form of cellulose, hemicelluloses, and lignin. Microorganisms have potential to degrade these organic, rich lignocellulosic wastes as a result of their metabolism and generate a metabolite having a fuel property along with various biochemicals (Figure 1.1). Though lignocellulosic biomass is available in abundance, it varies drastically in its composition.

Generally, lignocellulosic biomass is composed of 60%–80% complex carbohydrates of C5 and C6 sugar units and 20%–30% lignin by weight. The major limitations for sustainable production of biobased alcohols are the presence of lignin, recalcitrant characteristics of cellulose structure and efficient bioconversion processes and technologies. To overcome such limitations, competent pretreatment methods that can help to break down the recalcitrant structure are needed, making it more accessible to the enzymes.

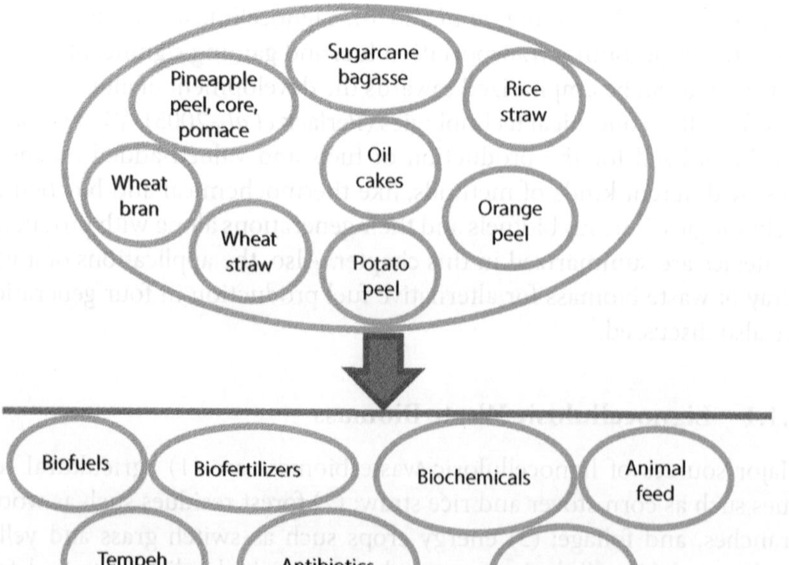

Figure 1.1 The production of various value-added products from agro-wastes through microbial fermentation.

1.1.2 Generations of Biofuels

Rapid global industrialization and population increase has witnessed a substantial hike in the production of municipal and industrial waste, having great potential feedstock for biofuel production (Bhatia *et al.*, 2020). These wastes can be recycled and reutilized to produce biofuels and biomaterials (Figures 1.2–1.4). Waste management practices have focused on the exploration of the potential of these wastes to curtail the effect of waste on the environment. Two major issues are associated when these waste materials are employed as feedstock for biofuel generation such as a clean source of energy and issues related to waste disposal. Using this approach, various liquid biofuels such as ethanol and biodiesel, biogas such as methane and CO_2, syngas (hydrogen and CO), or pure hydrogen are produced at various generation levels. There are many factors that govern the cost of biofuel production such as the efficiency of the process, scale of production, accessibility, and cost of feedstock (Bhatia *et al.*, 2019). Out of these, the feedstock price shares a major portion of the total cost input in biofuel production. Based on the generation types, there are four types of biofuels like first-generation biofuels, second-generation biofuels, third-generation biofuels, and fourth-generations biofuels (Figure 1.5).

Figure 1.2 Rice straw used in a rural area.

First-generation biofuels (Figure 1.6) are based on food crops like starch, grain, food grains, sugarcane juice, oats, barley, sugar crops, and oil seeds. Initially, such a generation showed potential in fossil fuels combustion, reducing atmospheric levels of CO_2 (Rodionova *et al.*, 2017). It, however, revealed a food security issue that limited the use of these feedstock for biofuel production. Various parameters are also responsible for evaluation of first-generation biofuels such as the biomass composition, availability of croplands, competition with food requirements, and cost of the biomass (Singh *et al.*, 2013). Long before, bioethanol was used in many industries in Germany and France during 1894 (Chelf *et al.*, 1993). During 1925, it was used as transportation fuel in Brazil. The potential of ethanol was detected as fuel in Europe and the United States until the 1900s, which was later stopped due to high production cost (Packer *et al.*, 2016). Since last three decades, bioethanol showed great attention as an alternative transportation fuel because countries like Brazil and the United States encouraged the domestic production of bioethanol.

Figure 1.3 Roadside rice straw heap.

Figure 1.4 Roadside sugarcane bagasse heap.

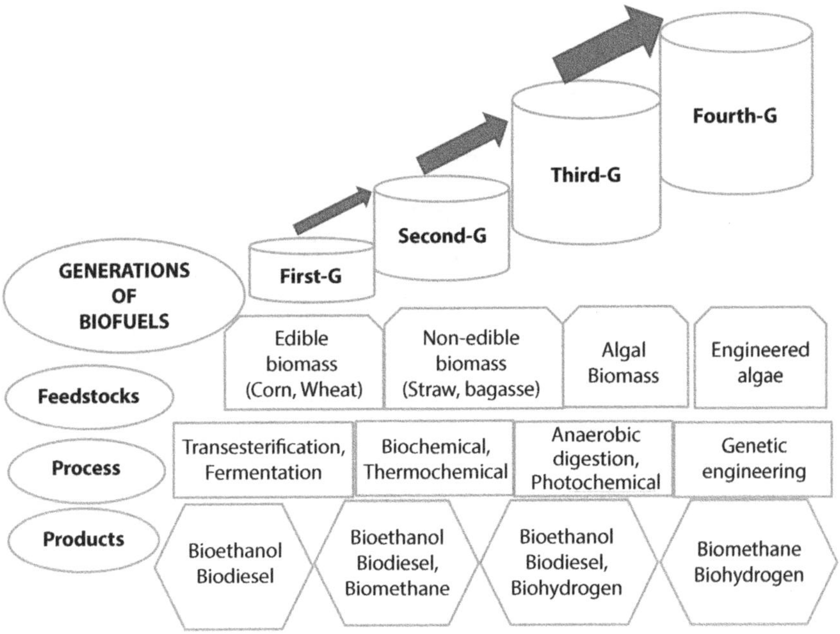

Figure 1.5 Generations of biofuels.

A recent study showed that the production of bioethanol is generally derived from food crops like sugarcane, maize, wheat, sugar beet, and sorghum, which can feed 200 million people (Rulli *et al.*, 2006), thereby creating a big debate about food versus fuel. Among food crops, sugarcane is the highest crop used in bioethanol production, requiring less water as compared to maize and wheat (Rulli *et al.*, 2006).

The first generation of biodiesel fuel similarly solely depended on oil crops, and it was detected that 75% of its production cost was dependent on the cost feedstock production (Mahmudul *et al.*, 2017). More than 350 oil-bearing crops with both edible and non-edible characteristics could be used as feedstock for biodiesel production. Common non-edible food crop sources that are used as sources for biodiesel production are rapeseed, soybean, palm, sunflower, peanut, safflower, corn, rice bran, coconut, olive, castor, milkweed seed, and linseed. These are *Jatropha curcas, Pongamia glabra, Madhuca indica, Salvadora oleoides*, cotton seed oil, tobacco, *Calophyllum Eruca*, Jojoba, neem oil, Moringa *oleifera* and *croton megalocarpus* (Alalwan *et al.*, 2019).

The first generation of biofuel is also based on bioethers (also known as fuel ethers), which can be used to increase the octane number of fuels, thereby

Figure 1.6 First-generation feedstock.

replacing petro-ethers and improving engine performance. Bioethers are produced with the reaction of bioethanol such as isobutylene. Additionally, bioethers can greatly reduce toxic exhaust emissions, having an important limitation of low-energy density. For the enhancement of transportation fuel quality, ether additives such as dimethyl ether (DME), methyl tertiary butyl ether (MTBE), diethyl ether (DEE), tert-amyl methyl ether (TAME), ethyl tert-butyl ether (ETBE), and tert-amyl ethyl ether (TAEE) are generally used. In Europe, MTBE and ETBE have been used for replacing highly toxic compounds like lead.

Several factors such as food security issues, climate change mitigation, and the usage of fertile lands restrict the production of first-generation biofuels, thereby shifting the focus towards utilizing nonfood feedstock like agro-waste (rice husk, groundnut shell, etc.), forest waste, wood processing residues, non-edible parts of sugar beet, sugarcane, and corn for the production of second-generation (2G) biofuels, showing no competition to the human food chain (Pathak and Chandel, 2017).

The second-generation biomass (Figure 1.7) is based on the non-edible plant residues having no food-versus-fuel issues that can be directly used in biorefineries for fuel production. Although feedstock need pretreatment methods for easy availability of fermentable sugars, thereby requiring additional processing steps and an operational cost for biofuel production. Vast amounts of agricultural crop residues are obtained due to agricultural practices such as cereals straw, corn cobs, cotton stalks, various grasses, maize and sorghum stover, sugarcane and tequila bagasse, coconut and banana residues, corn husks, coffee pulp and coffee husk, cottonseed and sunflower seed hulls, peanut shells, rice husks, sunflower seed hulls, waste paper, wood sawdust, and chips. These also include all agricultural wastes such as straw, stem, stalk, leaves, husk, shell, peel, lint, seed/stones, pulp, and stubble, which come from cereals (rice, wheat, maize or corn, sorghum, barley, millet), cotton, groundnut, jute, legumes (tomato, bean, soya) coffee, cacao, olive, tea, and fruits (banana, mango, coco, cashew). These are widely used as household fuel in developing countries. The forest residues include stems, barks, twigs, cones, and needles and are also used as the biomass for second-generation biofuel production. Wood processing also produces biomass as sawdust, wood chips, lumps, bales, pellets, and briquettes, which are utilized for biofuel production.

For the production of biodiesel, second-generation feedstock such as energy crops, agricultural residues, and wood wastes can be utilized. Generally used energy crops for the production of biodiesel are *Jatropha*, salmon oil, Rubber tree, *Aleurites moluccana*, *Madhuca longifolia*, and Jojoba oil. Additionally, various waste materials from food industries such as cooking oils, restaurant grease, beef tallow, animal fats, and pork lard can also be used for the production of biodiesel (Bhuiya *et al.*, 2016).

Figure 1.7 Second-generation feedstock.

Aquatic feedstock is used as the main source for third-generation biofuels (Gambelli *et al.*, 2017). Third-generation biofuels are based on algal sources, having a shorter generation time as compared to the other feedstock. Algal biomass can be utilized at its full potential, if the challenges associated with its cultivation, production process, design of bioreactor for algal production, and downstream processes are managed successfully. By the process of photosynthesis, high quantities of CO_2 were captured to produce O_2 and oil (Hayes *et al.*, 2015). Biofuels produced by such generations have low stability due to their being highly volatile at high temperatures thereby degrading easily (Shah *et al.*, 2018).

Microalgae (Figure 1.8) not only capture CO_2 from the atmosphere but also grow on polluted wastewater to absorb the environmental contaminants that can be considered as the biomass for third-generation biofuel (Rawat *et al.*, 2011). In general, algal biomass has the potential to fix the CO_2 to accumulate lipids and polysaccharides within their cells, thereby converting into various biofuels and other value-added industrial products. Algae are considered as one of the most capable biofuel feedstock due to their higher demand, photosynthetic efficiency, and less use of land or water (Renuka *et al.*, 2013). Additionally, algal biomass has great potential

Figure 1.8 Third-generation feedstock.

for third-generation biomass for their rapid growth rates and ability to absorb different nutrients in wastewater by utilizing CO_2 from the atmosphere or flue gas. It is detected that algal oil containing approximately 35,800 kJ/kg of energy has an equivalent to about 80% of the energy in petroleum (Iqbal *et al.*, 2017).

Finally, recent technologies-based fourth-generation biofuels deal with genetically modified microorganisms such as microalgae, yeast, fungi, and cyanobacteria to produce biofuels. Due to the wide array of applications of microalgae for the production of biofuels and value-added chemicals, the supplementation of genetic engineering field can spread the spectrum into wide ranges owing to improve the high yield and create an artificial carbon sink to eliminate or reduce carbon emission.

Cyanobacteria have been greatly involved in bioenergy and biofuel industries. Due to development of genetic engineering for several photosynthetic algae, the production of wide arranges of value-added products along with their process technologies were optimized. The most used eukaryotic microalgae are *Chlamydomonas reinhardtii*, which are used for experiment purposes for oxygenic photosynthesis, circadian rhythms, and flagella biogenesis. Other algae that are also experimented through genetic engineering process for production of biofuels and other industrially important chemicals are *Chlorella* and diatoms (Valenzuela *et al.*, 2018).

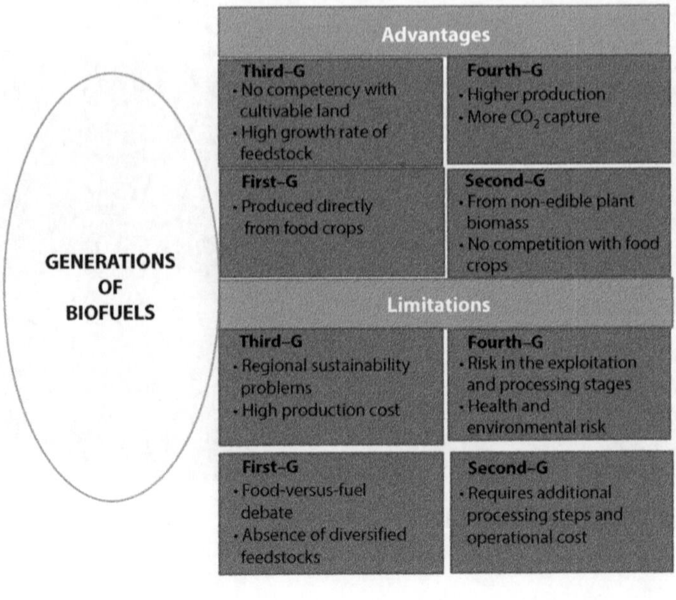

Figure 1.9 The advantages and disadvantages of different generations of biofuel.

The global production of biofuels is demanded by various parameters, of which feedstock is the most important one. Most disadvantages are detected in the first generation due to food-versus-fuel issues. Though other generations have more benefits to produce biofuels, there were still some bottlenecks during the processing, cost, and experimentations (Figure 1.9). Various biomolecules like carbohydrates and proteins; fibers and minerals are found in agro-wastes, which are the industrial interest. For an economic and environmental beneficial reutilization, these wastes are necessitated around the globe among researchers and academicians because of its economic feasibility that curtails the cost of production. Several value-added products can be produced by employing agro-forest waste and the waste generated out of food processing (Pleissner *et al.*, 2017). This approach will support the world's biobased economy, as well as resolve the issue of disposal.

1.1.3 Possibilities and Challenges

Molasses is the ultimate source of ethanol production in India and as per the targets set by the MNRE (Ministry of New Renewable Energy), sugarcane production has increased up to 20%–23% than that of its usual production. An area of about 10.5 million hectares may be required to grow 736.5 million tons of sugarcane, thereby depriving the cultivation of other food crops (Shinoj *et al.*, 2011; Ray, 2012). Lignocellulosic biomass. along with other organic matters, have great impact for ethanol production, lessening the dependency on molasses (Subramanian *et al.*, 2005). The second-generation process employs lignocellulosic material that has no direct connection with the food chain. These can be provided throughout the year. The global production of agricultural waste reached 1.4 billion tons annually (Saini *et al.*, 2015). As per the data, about 154–185 x 10^6 tons wheat straw and 317–380 x 10^6 tons of bagasse is generated per year worldwide (Carmen, 2009). About 179 metric tons of sugarcane bagasse is generated yearly in India, which makes India the second leading producer after Brazil (Kapoor *et al.*, 2006). As far as the production of fruits in India is concerned, about 75 x 10^6 tons (in 6383x10^3 ha of land), the overall waste generated comes to 25 million tons per year (India Agricultural Research Data Book 2013). Similarly, the production of pineapple also produces a lot of waste amounting to about 60%, which can be converted into biofuels and chemicals (Sarangi *et al.*, 2020a). India and China together share 91% of the world's lychee (*Litchi chinensis*) production. Other agro-wastes like peels of sweet oranges (*Citrus sinensis var mosambi*) and pineapples (*Ananas cosmosus)* are the ideal substrate for production of bioethanol because they do not compete with the food chain and have a high content of cellulose. Peels of *C. sinensis var mosambi* constitute a significant non-edible lignocellulosic biomass, which are generated widely all over India.

There are various advantages related to the usage of biofuels, making them an attractive and sustainable alternative option to replace fossil fuels. Furthermore, some socio-economic benefits can be manifested for rural development. Ethanol can serve the global hikes in energy demand along with its potential to reduce the emission of greenhouse gases of up to 85% (Perlack *et al.*, 2005). Among all biofuels, ethanol is one of the most dependable fuels, and hence, its global production has also gained huge success ranging from 14 to 26 million tons in recent times. Petrol can be blended with ethanol in various proportions, such as E10, E85, E95, containing 10%, 85% and 95% of ethanol, respectively. When 10% of ethanol is blended with petrol, it does not require any alterations in engine (Balat, 2009). Petroleum consumption is reduced after blending it with ethanol. Production of bioethanol by various generations is shown in Figure 1.10. The government of India introduced the Ethanol Blended Petrol Programme (EBPP) in January 2003 with an idea of merging petrol with ethanol. Also, the Ministry of New and Renewable Energy (MNRE)

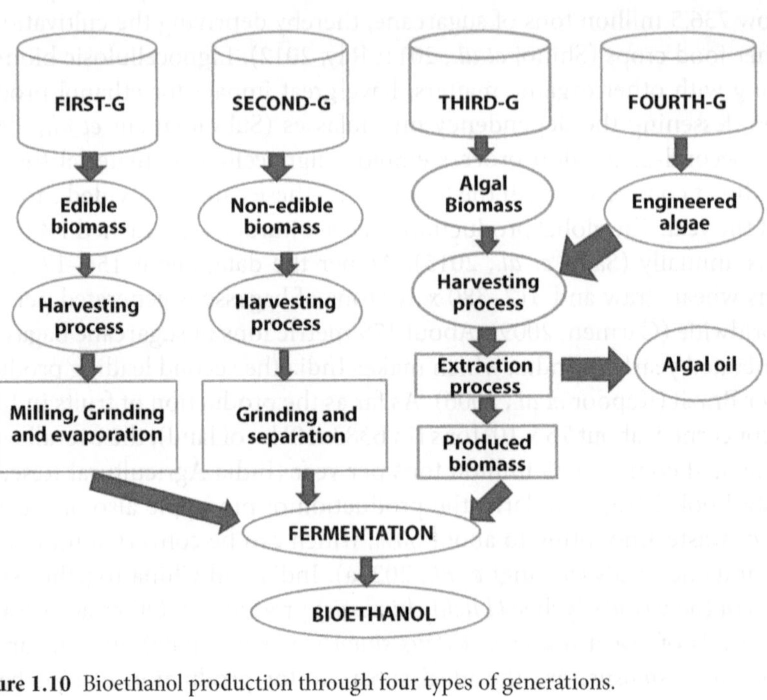

Figure 1.10 Bioethanol production through four types of generations.

developed the National Policy on Biofuels in 2009 regarding the blending of diesel and petrol with 20% biofuel by 2017. This target would demand 3.4 billion liters of bioethanol by 2020 (Basavaraj *et al.*, 2013).

1.2 Conclusion

Biofuels hold immense potential to completely or partially substitute petroleum-based fuels, thereby supporting the economy and the environment. In this chapter, the four generations of biofuels are described with respect to their feedstock, production technologies, and processing technologies. Also, the potential of various waste materials toward biofuel production is summarized. First-generation biofuel competes with food sources, thereby limiting its applications, whereas other generations have no such issues. In contrast, the third- and fourth-generation biofuels are more promising choices because they do not involve such food-versus-fuel issues. Prevailing infrastructure gives sufficient space to successfully incorporate biomass-based biofuels as transportation fuel. To avoid the debate between food and fuel, second-generation biofuels are mostly considered as compared to first-generation biofuels. Strategies based on the utilization of ubiquitous biomass feedstock and particularly non-edible lignocellulosic biomass would be more sustainable. As far as the reduction in greenhouse gas emissions is concerned, fourth-generation biofuels have the best potential. The second- and third-generation biofuels still have some challenges like high investment costs, as well as low conversion efficiencies of biomass. Further improvements in processing technologies of both the second and third generations will build a solid platform to produce bioethanol and biodiesels. A thriving lignocellulosic biorefinery can be established by synergizing various technologies and biomass-processing approaches for the supportive production of various fuel and chemical products. The advancement of microbial technology is a viable option to address many issues related to the commercialization of biofuels and other value-added products in the future.

References

Alalwan, H.A., Alminshid, A.H., Aljaafari, H.A.S., Promising evolution of biofuel generations. Subject review. *Renew. Energy Focus*, 28, 126–139, 2019.

Azkawi, A.S.A., Sivakumar, N., Bahry, S.A., Bioprocessing of cardboard waste for cellulase production. *Biomass Convers. Biorefin.*, 8, 597–606, 2018.

Balat, M., Global status of biomass energy use. *Energy Source A*, 31, 1160–1173, 2009.

Basavaraj, G., Rao, P.P., Basu, K., Reddy, R., Kumar, A.A., Rao, P.S., Reddy, B.V.S., Assessing viability of bio-ethanol production from sweet sorghum in India, 2013.

Bhatia, L., Sarangi, P.K., Nanda, S., Current Advancements in Microbial Fuel Cell Technologies, in: *Biorefinery of Alternative resources: Targeting green fuels and platform chemicals*, S. Nanda, D.V.N. Vo, P.K. Sarangi (Eds.), pp. 477–494, Springer Nature, 2020.

Bhatia, L., Sharma, A., Bachetti, R.K., Chandel, A.K., Lignocellulose derived functional oligosaccharides: production, properties and health benefits- A Review. *Process Biochem. Biotechnol.*, 49, 8, 744–758, 2019, https://doi.org/10.1080/10826068.2019.1608446.

Bhuiya, M.M.K., Rasul, M.G., Khan, M.M.K., Ashwath, N., Azad, A.K., Prospects of 2nd generation biodiesel as a sustainable fuel—Part: 1 selection of feedstocks, oil extraction techniques and conversion technologies. *Renew. Sustain. Energy Rev.*, 55, 1109–1128, 2016.

Byadgi, S.A. and Kalburgi, P.B., Production of bioethanol from waste newspaper. *Proc. Environ. Sci.*, 35, 555–562, 2016.

Hayes, C.J., Burgess Jr., D.R., Manion, J.A., Williams, I.H., Williams, N.H., Combustion pathways of biofuel model compounds: A review of recent research and current challenges pertaining to first-, second-, and third-generation biofuels, in: *Advances in Physical Organic Chemistry*, vol. 49, pp. 103–187, 2015.

Carmen, S., Lignocellulosic residues: Biodegradation and bioconversion by fungi. *Biotechnol. Adv.*, 27, 185–194, 2009.

Cea, M., Sangaletti-Gerhard, N., Acuna, P., Fuentes, I., Jorquera, M., Godoy, K., Osses, F., Navia, R., Screening transesterifiable lipid accumulating bacteria from sewage sludge for biodiesel production. *Biotechnol. Rep*, 8, 116–123, 2015.

Chandel, A.K., Antunes, F.A.F., Hilares, R.T., Cota, J., Ellila, S., Silveira, M.H.L., da Silva, S.S., Bioconversion of Hemicellulose into Ethanol and Value-Added Products: Commercialization, Trends, and Future Opportunities, in: *Advances in Sugarcane Biorefinery*, 2018a.

Chandel, K., Garlapati, V.K., Singh, A.K., Antunes, F.A.F., da Silva, S.S., The path forward for lignocellulose biorefineries: Bottlenecks, solutions, and perspective on commercialization. *Bioresour. Technol.*, 264, 370–381, 2018b.

Cheah, W.Y., Ling, T.C., Juan, J.C., Lee, D.J., Chang, J.S., Show, P.L., Biorefineries of carbon dioxide: from carbon capture and storage (CCS) to bioenergies production. *Bioresour. Technol.*, 215, 346–356, 2016a.

Chelf, P., Brown, L.M., Wyman, C.E., Aquatic biomass resources and carbon dioxide trapping. *Biomass Bioenergy*, 4, 175–183, 1993.

Cobucci-Ponzano, B., Strazzulli, A., Iacono, R., Masturzo, G., Giglio, R., Rossi, M., Moracci, M., Novel thermophilic hemicellulases for the conversion of lignocellulose for second generation biorefineries. *Enzyme Microb. Technol.*, 78, 63–73, 2015. https://doi.org/10.1016/j.enzmictec.2015.06. 014.

Dumon, C., Song, L., Bozonnet, S., Fauré, R., O'Donohue, M.J., Progress and future prospects for pentose-specific biocatalysts in biorefining. *Process Biochem.*, 47, 346–357, 2012. https://doi.org/10.1 016/j.procbio.2011.06.017.

El-Tayeb, T.S., Abdelhafez, A.A., Ali, S.H., Ramadan, E.M., Effect of acid hydrolysis and fungal biotreatment on agro-industrial wastes for obtainment of free sugars for bioethanol production. *Braz. J. Microbiol.*, 43, 4, 1523–1535, 2012.

Escobar-Niño, A., Luna, C., Luna, D., Marcos, A.T., Ca´novas, D., Selection and Characterization of Biofuel-Producing Environmental Bacteria Isolated from Vegetable Oil-Rich Wastes. *PLoS One*, 9, 8, e104063, 2014.

Hamedi, J. and Azimi, A., Identification and Prioritization of Interorganizational Success Factors in Microbial Biotechnology Firms. *J. Virol. Microbiol.*, 2013, 190528, 8, 2016. DOI: 10.5171/2013.190528.

Iqbal, S.Z., Malik, N., *et al.*, Processing techniques of algae-based materials, in: *Algae Based Polymers, Blends, and Composites*, K.M. Zia, M. Zuber, M. Ali (Eds.), pp. 671–686, Elsevier, Netherlands, 2017.

Kapoor, R.K. and Chandel, A.K., Bioethanol from crop residues, production forecasting and economics: An Indian perspective, in: *Lignocellulose Biotechnology: Current and Future Prospects*, R.C. Kuhad and A. Singh (Eds.), pp. 32–45, I.K. International, New Delhi, 2006.

Kaur, I. and Sahni, G., Multi-Scale Structural Studies of Sequential Ionic Liquids and Alkali Pretreated Corn Stover and Sugarcane Bagasse. *Green Sustain. Chem.*, 8, 92–114, 2018.

Ballesteros, L.F., Teixeira, J.A., Mussatto, S.I., Chemical, functional, and structural properties of spent cofee grounds and coffee silver skin. *Food Bioprocess Technol.*, 7, 3493–3503, 2014.

Lun, L.W., Gunny, A.A.N., Kasim, F.H., Arbain, D., Fourier transform infrared spectroscopy (FTIR) analysis of paddy straw pulp treated using deep eutectic solvent. *AIP Conference Proceedings*, vol. 1835, p. 020049, 2017, doi: 10.1063/1.4981871.

Mahmudul, H.M., Hagos, F.Y., Mamat, R., Adam, A.A., Ishak, W.F.W., Alenezi, R., Title ???. *Renew. Sustain. Energy Rev.*, 72, 497–509, 2017.

Martin, J.G.P., Porto, E., Correa, C.B., Alencar, S.M., Gloria, E.M., Cabral, I.S.R., Aquino, L.M., Antimicrobial potential and chemical composition of agroindustrial wastes. *J. Nat. Prod.*, 5, 27–36, 2012.

Motte, J.C., Trably, E., Escudié, R., Hamelin, J., Steyer, J.P., Bernet, N., Delgenes, J.P., Dumas, C., Total solids content: a key parameter of metabolic pathways in dry anaerobic digestion. *Biotechnol. Biofuels*, 6, 164, 2013.

Nanda, S. and Rana, R., *et al.*, A Spotlight on Butanol and Propanol as Next-Generation Synthetic Fuels, in: *Biorefinery of Alternative resources: Targeting green fuels and platform chemicals*, S. Nanda, D.V.N. Vo, P.K. Sarangi (Eds.), pp. 105–126, Springer Nature, 2020, ISBN: 978-9811518034.

National Policy on Biofuels Ministry of New and Renewable Energy (MNRE), Government of India, 2009.

Packer, M.A., Harris, G.C., Adams, S.L., Food and Feed Applications of Algae, in: *Algae Biotechnology: Products and Processes*, Bux, F. and Chisti, Y. (Eds.), pp. 217–247, 2016.

Pasin, T.M., de Almeida, P.Z., *et al.*, Bioconversion of Agro-industrial Residues to Second-Generation Bioethanol, in: *Biorefinery of Alternative Resources: Targeting Green Fuels and Platform Chemicals*, S. Nanda, D.V.N. Vo, P.K. Sarangi (Eds.), Springer, Singapore, 2020.

Pathak, B.S. and Chandel, A.K., Feedstock Transportation, Agricultural Processing, Logistic from Farm to Bio-refinery: Recent Developments, Mechanization, and Cost Analysis, in: *Sustainable Biofuels Development in India*, A.K. Chandel and R.K. Sukumaran (Eds.), Springer International Publishing, 2017.

Perlack, R.D., Wright, L.L., *et al.*, Biomass as feedstock for a bioenergy and bioproducts industry: the technical feasibility of a billion-ton annual supply. DOE/GO-10 2005-2135, April. U.S. Department of Energy and U.S. Department of Agriculture, Oak Ridge National laboratory, USA, pp. 1–78, 2005.

Philippini, R.R., Martiniano, S.E., Chandel, A.K., de Carvalho, W., da Silva, S.S., Pretreatment of Sugarcane Bagasse from Cane Hybrids: Effects on Chemical Composition and 2G Sugars Recovery. *Waste Biomass Valorization*, 2017.

Pleissner, D., Demichelis, F., Mariano, S., Fiore, S., Guitierrez, I.M.N., Schneider, R., Venus, J., Direct production of lactic acid based on simultaneous saccharification and fermentation of mixed restaurant food waste. *J. Clean. Prod.*, 143, 615–623, 2017.

Rawat, I., Kumar, R.R., Mutanda, T., Bux, F., Dual role of microalgae: phycoremediation of domestic wastewater and biomass production for sustainable biofuels production. *Appl. Energy*, 88, 3411–3424, 2011.

Ray, S., Goldar, A., Miglani, S., The Ethanol Blending Policy in India. *Econ. Political Wkly*, 1, 23–25, 2012.

Renuka, N., Sood, A., Ratha, S.K., Prasanna, R., Ahluwalia, A.S., Evaluation of microalgal consortia for treatment of primary treated sewage effluent and biomass production. *J. Appl. Phycol.*, 25, 1529–1537, 2013.

Rivas, B., Torrado, A., Torre, P., Converti, A., Domínguez, J.M., Submerged citric acid fermentation on orange peel autohydrolysate. *J. Agric. Food. Chem.*, 56, 2380–2387, 2008.

Rodionova, M., Poudyal, R., Tiwari, I., Voloshin, R., Zharmukhamedov, S., Nam, H., Zayadan, B., Bruce, B., Hou, H., Allakhverdiev, S., Biofuel production: Challenges and opportunities. *Int. J. Hydrogen Energy*, 42, 8450–8461, 2017.

Rulli, M., Bellomi, D., Cazzoli, A. *et al.*, The water-land-food nexus of first-generation biofuels. *Sci Rep* 6, 22521. 2016.. https://doi.org/10.1038/srep22521

Saini, J.K., Saini, R., Tewari, L., Lignocellulosic agriculture wastes as biomass feedstocks for second-generation bioethanol production: concepts and recent developments. *Biotech.*, 5, 4, 337–353, 2015.

Sarangi, P.K., Singh, J., Singh, T.A., Pectin from pineapple waste: Isolation and Process Optimization. *Int. J. Curr. Microbiol. Appl. Sci.*, 9, 143-148, 5, 2020b.

Sarangi, P.K., Singh, J., Singh, T.A., Potential of pineapple wastes for Pectin recovery: A path towards waste valorization. *Int. J. Chem. Stud.*, 8, 3, 1212–1214, 2020a.

Sarangi, P.K. and Nanda, S., Bioconversion of agro-wastes into phenolic flavour compounds, in: *Biotechnology for Sustainable Energy and Products*, P.K. Sarangi and S. Nanda (Eds.), pp. 266–284, IK International Publishing House, India, 2019a.

Sarangi, P.K. and Nanda, S., Valorization of pineapple wastes for biomethane generation, in: *Biogas Technology*, S. Mishra, T.K. Adhya, S.K. Ojha (Eds.), pp. 169–180, New India Publishing Agency, New Delhi, India, 2019b.

Sarangi, P.K. and Nanda, S., Recent developments and challenges of acetone-butanol-ethanol fermentation, in: *Recent advancements in biofuels and bioenergy utilization*, P.K. Sarangi, S. Nanda, P. Mohanty (Eds.), pp. 111–123, Springer Nature, Singapore, 2018.

Sarangi, P.K. and Nanda, S., Recent advances in consolidated bioprocessing for microbe-assisted biofuel production, in: *Fuel processing and energy utilization*, S. Nanda, P.K. Sarangi, D.V.N. Vo (Eds.), pp. 141–157, CRC Press, Boca Raton, FL, 2019c.

Sarangi, P.K. and Nanda, S., Biohydrogen Production through Dark Fermentation. *Chem. Eng. Technol.*, 43, 4, 601–612, 2020.

Sarangi, P.K., Nanda, S., Sahoo, H.P., Maximization of vanillin production by standardizing different cultural conditions for ferulic acid degradation. *N. Y. Sci. J.*, 3, 7, 77–79, 2010.

Sarangi, P.K., Nanda, S., Vo, D.V.N., Technological Advancements in the Production and Application of Biomethanol, in: *Biorefinery of Alternative resources: Targeting green fuels and platform chemicals*, S. Nanda, D.V.N. Vo, P.K. Sarangi (Eds.), pp. 127–140, Springer Nature, 2020c.

Sarangi, P.K., Sahoo, H.P., Ghosh, S., Mitra, A., Biotransformation of ferulic acid into vanillin by a *Streptomyces* isolate S10. *Asian J. Microbiol. Biotechnol. Environ. Sci.*, 11, 273–278, 2009b.

Sarangi, P.K. and Sahoo, H.P., Biotransformation of ferulic acid into value-added products. *Asian J. Microbiol. Biotechnol. Environ. Sci.*, 12, 3, 523–525, 2010a.

Sarangi, P.K., Sahoo, H.P., Sci., J., Ferulic Acid production from wheat bran using *Staphylococcus aureus*. *N. Y., Sci. J.*, 3, 4, 79–81, 2010b.

Sarangi, P.K. and Sahoo, H.P., Enhancing the rate of ferulic acid bioconversion utilizing glucose as carbon source. *J. Am. Sci.*, 6, 5, 115–117, 2010c.

Sarangi, P.K., Sahoo, H.P., Ghosh, S., Mitra, A., Isolation of ferulic acid from wheat bran using *Streptomyces* isolates. *Int. J. Appl. Agric. Sci.*, 4, 1, 57–62, 2009a.

Sarangi, P.K., Singh, T.A., Ng., J., Pineapple as potential crop resource: Perspective and value addition, in: *Food Bioresources and Ethnic Foods of Manipur, Northeast, India*, pp. 83–91, Empyreal Publishing House, 2019.

Sarks, C., Jin, M., Sato, T.K., Balan, V., Dale, B.E., Studying the rapid bioconversion of lignocellulosic sugars into ethanol using high cell density fermentations with cell recycle. *Biotechnol. Biofuels*, 7, 73, 2014. https://doi.org/10.1186/1754-6834-7-73.

Shinoj, P., Raju, S.S., Joshi, P.K., India's biofuels production programme; need for prioritizing alternative options. *Indian J. Agric. Sci.*, 81, 5, 391–397, 2011.

Siang, T.J., Roslan, N.A., *et al.*, Recent Advances in Steam Reforming of Glycerol for Syngas Production, in: *Biorefinery of Alternative resources: Targeting green fuels and platform chemicals*, vol. ??, S. Nanda, D.V.N. Vo, P.K. Sarangi (Eds.), pp. 399–426, Springer Nature, 2020.

Singh, A., Bajara, S., Bishnoia, N.R., Physico-chemical pretreatment and enzymatic hydrolysis of cotton stalk for ethanol production by *Saccharomyces cerevisiae*. *Bioresour. Technol.*, 244, 71–77, 2017.

Singh, A., Olsen, S., II, Pant, D., Importance of life cycle assessment of renewable energy sources, in: *Life Cycle Assessment of Renewable Energy Sources*, vol. ???, pp. 1–11, Springer, 2013.

Subramanian, K.A., Singal, S.K., Saxena, M., Singal, S., Utilization of liquid biofuels in automotive diesel engines: An Indian perspective. *Biomass Bioenergy*, 29, 65–72, 2005.

United States Energy Information Administration, USEIA, International Energy Outlook 2011, 2011, Accessed from: http://www.eia.gov/forecasts/ieo/pdf/0484(2011).pdf. Accessed 3 Jan 2012.

2

Pretreatment Methods of Waste Biomass: A Broad Overview

Prakash Kumar Sarangi

College of Agriculture, Central Agricultural University, Imphal, Manipur, India

Abstract

The production of bioethanol is a multistep process wherein steps to the pretreatment of lignocellulosic biomass is pivotal. Several strategies are now available for pretreating biomass. The pretreatment process can either be physical, chemical, physicochemical, or biological. Merits and demerits are associated with each pretreatment approach. Enzymatic saccharification becomes easier and efficient once the biomass is pretreated. Alteration in structure of pretreated biomass allows easier accession towards their enzymatic hydrolysis, thereby enhancing the yield of bioethanol. This chapter discusses various novel approaches of pretreatment like supercritical fluid pretreatment and hydrodynamic cavitation, along with conventional approaches like acid, alkali, wet oxidation, ionic liquid, ammonia recycle percolation, and steam explosion.

***Keywords*:** Steam explosion, crystallinity, supercritical fluid pretreatment, hydrodynamic cavitation, enzymatic hydrolysis, biological pretreatments, supercritical CO_2 ($scCO_2$)

2.1 Introduction

The world heavily relies upon fossil fuels as a primary source of energy. Rapid civilization, unprecedented industrialization, continued population growth, and an ever-expanding economy are the factors behind the hike in energy demand and depletion of natural energy sources globally (Pathak and Chandel, 2017). The consumption of fossil fuel is predicted to hike from 15 TW in 2012, to two times or three times the bulk in 2050 and 2100, mutually on a global level (Cheah *et al.*, 2018). This planet may

Prakash Kumar Sarangi. Waste Biomass to Alternative Fuels, (21–44) © 2025 Scrivener Publishing LLC

face extreme challenges, such as a shortage of fossil fuels along with the rise in global warming and pollution. To circumvent these challenges, it is high time to explore the potential of renewable sources of energy like solar, wind, hydro, tidal, and biomass globally (Pathak and Chandel, 2017). These novel alternative sources will not be only sustainable but also cost effective. There are other substantial reasons to undergo into a massive-scale substitution of petroleum-based fuels, such as energy security, concomitant hikes in prices of crude oil, and unavoidable changes in climate (Goel and Wati, 2016).

Many countries are taking the initiative to address various issues of hikes in global energy demand and consumption (Cortes–Tolalpa *et al.*, 2016). The European Union is all set to use renewable sources of energy to reduce the need on imports of energy and enhance inventory guarantee (Cucchiella *et al.*, 2014). Recent research has diverted their attention towards generating energy resources that are not only alternative but also eco-friendly (Waghmare *et al.*, 2018). The exploration of sustainable fuel sources and essential chemicals from bio-resources is becoming more popular (Elmore *et al.*, 2017). Fossil fuel is presumed to be substituted by biomass-based renewable energy by 10%–50% in 2030 (Cucchiella *et al.*, 2014).

There are various forms of bioenergy. Biofuel is one of them. It is a product of microbial activity on agricultural waste products, making it one of the potential sources of alternate energy (Wang *et al.*, 2014). Precious biochemicals are the outcome of various efforts made to alter microorganisms, by plugging-in and -out genes required for their production. Profit generated as an outcome of these artificial biological tools and efforts is synthesis of biofuel (Bhatia and Johri, 2017). Priorities have been diverted towards enhancing biofuel production from renewable lignocellulosic biomaterial, to address the pivotal issues of hike in global demand for energy and environmental threats in the form of global warming synergized with fossil fuels combustion (Liu *et al.*, 2015).

Biofuels are a form of bioenergy that are biologically produced from bio-organic matter. Biofuels can be solid, liquid, or gaseous. Based on the sources from where they are derived, biofuels can be classified into three categories: (1) Biofuels of natural origin; (2) Primary biofuels; and (3) Secondary biofuels. Landfill gas, animal waste, and vegetables are the organic sources of natural biofuels. Heating, cooking, electricity production, or brick furnace employs fuelwoods that are primary biofuels. Bioethanol, biobutanol, biodiesel, and biohydrogen are examples of secondary biofuels that are the outcome of biomass processing. Secondary biofuels can be of first generation or second generation. Food crops and oil seeds are the source of first-generation biofuels. Several factors restrict the production of first-generation biofuels; the biggest among them are food security issues, climate change mitigation, and the usage of fertile lands for fuel production. Therefore, the focus has shifted towards utilizing non-food feedstocks like agro-waste (rice husk, groundnut shell, etc.), forest waste, wood processing residues, non-edible parts of sugar beet, sugarcane, and corn for the production of second generation biofuels, which gives no competition to human food chain (Pathak and Chandel, 2017). Ethanol has gathered peak attention among all biofuels, and its consumption comprises more than 90% of the complete bioalcohols in the US (Reginatto and Antonio, 2015).

2.2 Broad Overview of Pretreatment

Chapter 2 thoroughly discusses various steps involved in ethanol production. Varieties of lignocellulosic biomass (LB) explored for ethanol production have been mentioned. These biomasses are largely available globally in nature, acting as a potent alternative to energy sources. The concept of biorefinery exists because of a broad spectrum of biofuels, biochemicals, and biomaterials produced in a sustainable manner from these biomasses. Complex LB cannot be directly fermented to ethanol by microorganisms, and hence demands a suitable pretreatment strategy, which aids in the hydrolysis of lignocelluloses into fermentable sugars.

There is significant involvement of microbes and/or their enzyme systems in ethanol production. The natural recalcitrance of LB creates hurdles for microbes and/or their enzyme systems to convert cellulosic and/or hemicellulosic sugars to ethanol via fermentation. This problem can be substantially solved by pretreating the LB that not only accelerated the hydrolysis but also enhanced the product yield. Physical and chemical barriers are removed once the pretreatment is done. These barriers were initially responsible for recalcitrance and inaccessibility towards enzymatic hydrolysis. During pretreatment, hemicelluloses and/or lignin gets solubilized, thereby making cellulose accessible for enzymatic hydrolysis (Jönsson and Martin, 2016). Accessibility of cellulolytic enzymes enhances after pretreatment as an outcome of the alteration in pore size of biomass and the reduction of cellulose crystallinity. As the accessibility of enzymes gets quicker and easier, less quantity of enzymes and therefore less cost input to manifest hydrolysis of pretreated biomass are seen as compared to unpretreated biomass (Dawson and Boopathy, 2007). Pretreatment also supports the upgradation of biodegradable materials (Yang and Wyman, 2008).

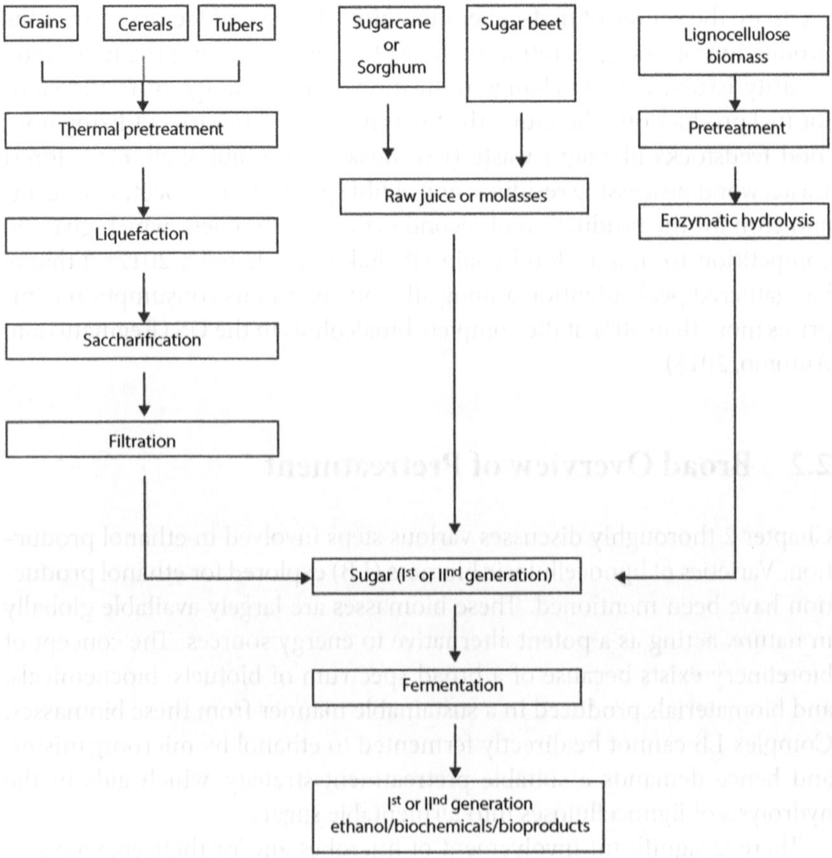

Figure 2.1 Basic technology and the main steps for bioethanol production from LB and starch-based feedstock.

As depicted in Figure 2.1, pretreatment is mostly the first step in producing second-generation sugars, which will later produce second-generation ethanol from LB and starch-based feedstock.

Several factors have been explored that play a pivotal role in the development of ideal cost-efficient and energy-efficient pretreatment procedures (Maurya *et al.*, 2015). There are many pretreatment strategies established to elevate the reactivity of cellulose and to improve the production of fermentable sugars. Important aims of pretreatment embrace:

i. Enhancing sugar production during enzyme hydrolysis of highly digestible solids produced post pretreatment;

ii. Prevention of degradation of sugars like pentoses and other sugars obtained from hemicelluloses;
iii. Curtailing the formation of inhibitors for the smooth conduction of subsequent fermentation steps;
iv. Lignin recovery so that it can get converted into valuable coproducts; and
v. Involving reactors of moderate size so that during operation minimized heat and power is required, thereby making the process cost effective (Yang and Wyman, 2008).

Pretreatment must face major challenges to accomplish above aims. These challenges include:

i. Establishing an accomplished technology that serves the purpose of achieving a significant volume of pretreated biomass material irrespective of the raw material employed;
ii. Achieving cost effective and eco-friendly methods that are able to perform best under ambient conditions;
iii. Reducing the initial investments, by employing cost effective construction material like steel alloys that can tolerate acid or base; and
iv. Involving fewer corrosive chemicals (Valdivia *et al.*, 2016).

Many forms of technological processes are now available that support the pretreatment of LB. Chief among these technological processes include chemical, physical, and biological processes. Some of these technologies have been successfully employed commercially, whereas a few are still being experimented on a lab scale.

Different pretreatment approaches lead to alterations in morphology and composition of biomass, leading to a reduction in its recalcitrance. For instance, biomass that has undergone steam explosion pretreatment has a disorganized structure with loose fibrous network. Cell wall components also gets solubilized. These changes allow easy accessibility of cell walls towards cellulase (Auxenfans *et al.*, 2017). Effective removal of lignin and/or hemicellulose can be gained via chemical pretreatment as alkaline pretreatment was given to sugarcane bagasse prior to its enzymatic saccharification (Terán–Hilares *et al.*, 2017a). The research group treated bagasse with 0.3 M NaOH solution at 70°C for 4 h that effectively removed 45% of lignin along with 49.2% and 57.0% respective digestion of hemicellulose and cellulose. This process effectively removed approximately 32.2% hemicellulose along with lignin. It was reported that after Na_2CO_3 pretreatment, crystallinity hiked about 25%in corn stover, 21% in switchgrass, and 15% in Miscanthus (Mirmohamadsadegh *et al.*, 2016). The retention of cellulose, partial solubilization of hemicelluloses, and removal of lignin (40%–59%) were the causes for this hike in crystallinity, respectively.

The rise in crystallinity is obvious as the amorphous fraction of biomass can be removed with ease. Enzymatic digestibility of cellulose gets enhanced as crystallinity of residual cellulose is lower when related to untreated biomass (Driemeier *et al.*, 2011). The development of new technical process optimization and conditions take many aspects into consideration. Chief among them includes understanding the variable interaction (e.g.: removal of lignin and increase of surface area), characteristics of biomass, and their changes during pretreatment.

Generally, the crystallinity index (CI) is a popular method to authenticate the changes in crystallinity of biomass due to its pretreatment. The effectiveness of this method has been questioned by some researchers, as this method cannot differentiate easily the definite crystallinity of the cellulose and overall biomass (Driemeier *et al.*, 2011). The researchers reported that when sugarcane bagasse was pretreated by dilute acid, soda delignification, hydrothermal, or steam explosion methods, evolution of cellulose crystals took place. It is presumed that non-crystalline cellulose is preferentially removed after pretreatment, so crystal-to-cellulose ratio should be high, whereas these researchers found that this ration was decreased. An alternative was provided to examine the effects of pretreatment. The application of cross polarization was demonstrated by multiple contact periods (Multi-CP) to obtain quantitative 13°C solid-state nuclear magnetic resonance (SSNMR) spectra to assess raw and pretreated sugarcane bagasse (Bernardinelli *et al.*, 2015). This was a better and a feasible approach to unfold action of different pretreatments in digestion of cell wall altering ultrastructure of cellulose.

2.2.1 Existing Physicochemical Promising Approaches for Pretreating Biomass

Even though numerous options of biomass pretreatment are available, its industrial implication still demands the continuous exploration of novel methods of cost-effective approaches, where new substitutes have been reported. Some of these approaches are described below.

2.2.1.1 *Physical Pretreatment Approaches*

Physical pretreatments are a cost-effective and eco-friendly approach of pretreating varieties of biomass with limited success. Ball milling and grinding are examples of physical pretreatment. There is an enormous possibility in these pretreatment approaches, as these approaches have been less explored despite being cost-effective and eco-friendly. Amalgamation of chipping, grinding and milling aids in comminution of waste biomass, thereby reducing cellulose crystallinity. The size of the materials obtained after chipping and milling/grinding is typically in the array of 10–30 mm and 0.2–2 mm, individually. Vibratory ball milling is a better approach than conventional ball milling, as the former has been found efficient in enhancing the degradability of the biomass feedstocks like spruce and aspen chips, as an outcome of changes in cellulose crystallinity (Millet *et al.*, 1976). The mechanical comminution of LB is a power intensive process. The requirement of power is decided by the characteristics of LB and the final particle size (Cadoche and Lopez, 1989). Pyrolysis also supports the pretreatment of LB (Shafizadeh and Bradbury, 1979). Ultrasound is identified as an effective approach of processing vegetal substances (Vinatoru *et al.*, 1999). Obtaining clean cellulose fiber from wastepaper was made possible when ultrasound was employed (Scott and Gerber, 1995). Ultrasound aids in saccharification processes (Rolz, 1986), along with the improvement in extraction of hemicellulose (Ebringerova and Hromadkova, 2002), cellulose (Pappas *et al.* 2002) and lignin (Sun and Tomkinson, 2002). The need for cellulase was curtailed by one third to half when the blended wastepaper was sonicated. Bioethanol production was also boosted to about 20% due to this approach (Wood *et al.*, 1997). The period involved in ultrasonic treatment is inversely proportional to the irradiation power applied (Imai *et al.*, 2004).

2.2.1.2 *Chemical Pretreatment*

A wide spectrum of acids, alkali, and oxidizing agents are involved in the chemical pretreatment of LB. The effect of pretreatment on structural components varies differ, based on the type of chemical employed. The removal of lignin becomes effective when pretreatments like alkaline, ozonolysis, peroxide, and wet oxidation pretreatments are given to the biomass. Hemicelluloses solubilization efficiently takes place on dilute acid pretreatment (Antunes *et al.*, 2019).

2.2.1.2.1 Acid Pretreatment Methods

Acid pretreatment is an efficient approach in harvesting a high yield of sugars from LB. The solubilization of hemicellulosic fraction results in easy accessibility of cellulases towards cellulose. Acids can be used in high concentration (72% w/w H_2SO_4) or in diluted form (0.4%–4% w/w H_2SO_4) at elevated temperature (120°C–200°C). This approach facilitates the digestion of LB efficiently as the tough structure of the lignocellulosic material gets damaged. Capital investments and the functioning cost of this pretreatment method are very high due to the involvement of high temperature and special reactor metallurgy. However, less enzyme loading is needed during enzymatic saccharification, as hemicellulose and cellulose gets partially depolymerized during acidic conditions. Sometimes, succeeding in enzyme-assisted hydrolysis stage is infrequently not required as fermentable sugars are generated during pretreatment itself (Zhu *et al.*, 2009). The cost of acids along with the necessity to detoxify the hydrolysates adds further cost in this strategy of pretreatment. The employment of concentrated acid is toxic, hazardous, and corrosive, thereby involving expensive reactors that are corrosion-resistant.

Switchgrass, corn stover, spruce (softwood), and poplar are some lignocellulosic biomasses that have been successfully pretreated with dilute H_2SO_4 (Taherzadeh and Karimi, 2008). Though more efficient than dilute sulfuric acid, concentrated sulfuric acid generates a variety of inhibitory substances like furfural (Goldstein and Easter, 1992). Decreased acid concentrations are required in dilute sulfuric acid pretreatment. However, it requires elevated temperatures and once more leads to furfural development.

A combination of sulfuric acid with acetic acid is directed to 90% saccharification (De Moraes–Rocha *et al.*, 2010). Efficiency of Organic acids like maleic acid and fumaric acid to pretreat LB has also been explored. The formation of furfural is significantly reduced to about three and five times as compared to that of dilute H_2SO_4 pretreatment (Onoghwarite *et al.*, 2016).

Ethanol production from corn and dilute acid pretreated corn stover (CS) as mixed substrates (Yu *et al.*, 2019). Dilute acid hydrolysis of sugarcane bagasse is an efficient approach to solubilize hemicellulose, thereby supporting the accessibility of cellulase to the remaining cellulignin matrix in the biomass (Philippini *et al.*, 2017). This approach could eliminate about 77% of hemicelluloses and 58% of lignin. About 80% of cellulose found in cellulose-rich pulp samples are left over. This pulp on further enzymatic hydrolysis gave a maximum saccharification yields of 55.39%.

As mentioned earlier, acid treatment also generates toxic substances that are inhibitory for microorganisms towards bioethanol formation. They also affect the viability yield of bioethanol and efficiency (Chandel *et al.*, 2007). A few examples of these toxic substances are furfural, 5-hydroxymethyl-furfural (5-HMF), acetic acid, formic acid, and levulinic acids (Parawira and Tekere, 2011). These substances can affect the rate of sugar uptake by microorganisms, thereby hindering their growth. The role of biological membranes also get disturbed because of these inhibitors (Chandel *et al.*, 2013). Hence, successive strategies should be adopted to detoxify these inhibitory substances. These strategies include the use of ion exchange resins or active charcoal. Enzymatic detoxification can be manifested utilizing laccase. Overliming with $Ca(OH)_2$ is also an effective approach (Fonseca *et al.*, 2011).

Inhibitors generated after pretreatment adversely affected the growth and metabolism of an organism due to a disturbance in the functioning of plasma membranes and other biological membranes. Incubation time gets prolonged, and a generation of metabolites also get reduced. Yield content might remain unaltered (Chandel *et al.*, 2013). Recent years have witnessed the employment of various detoxification strategies employing enzymes like laccases, the use of active charcoal, the ion exchange resins, and overliming with calcium hydroxide (Fonseca *et al.*, 2011).

2.2.1.2.2 Alkaline Pretreatment

Bases that are involved in the alkaline pretreatment of lignocellulosic pretreatment include NaOH, KOH, $Ca(OH)_2$, as well as NH_4OH (Aswathy *et al.*, 2010). Assembly of lignin gets damaged when biomass is treated with sodium hydroxide, thereby making it easier for cellulase and hemicellulase to act upon (Zhao *et al.*, 2008). The use of calcium hydroxide is a better approach over the use of NaOH, as the former is cost-effective, easily recoverable, and safe (Mosier *et al.*, 2005). Some of the effects of alkaline pretreatment include changes in lignin structure, distension of cellulose, and limited solvation of hemicellulosic materials (Sills and Gossett, 2011). Cellulosic material also gets decrystallized to some extent (Ibrahim *et al.*, 2011). These effects are manifested due to the breakdown of ester and glycosidic side chains along with separations of structural relations between lignin and carbohydrates. The internal surface area increases during this pretreatment along with the reduction in the degree of polymerization and crystallinity. Alkaline pretreatment takes place under mild and ambient physical conditions. These conditions make alkaline pretreatment a slow process.

2.2.1.2.3 Wet Oxidation

Wet oxidation has been explored to pretreat a variety of biomasses that include wheat straw, corn stover, sugarcane bagasse, cassava, peanuts, rye, canola, faba beans, and reed. Sugars like glucose and xylose were recovered upon successive enzymatic hydrolysis of these substrates (Ruffell *et al.*, 2010). These biomasses get oxidized when dissolved in water during their pretreatment. Water and biomass ratio is respectively 1:6. The size of LB is 2 mm as a result of its drying and milling, a practice that takes place prior to wet oxidation. Dissolution of hemicellulosic material along with the elimination of lignin content takes place in this process. Carbon dioxide, water, and carboxylic acids are formed because of lignin decomposition during the process. Wet oxidation aids in removal of dense wax covering bearing silica (Schmidt *et al.*, 2002). Wet oxidation is associated with alkaline to hydrolyze wheat straw (20 g straw per liter, 170°C, 5–10 min) (Bjerre *et al.*, 1996). In this process, 85% of cellulose got converted to glucose.

2.2.1.2.4 Ionic Liquids

The employment of ionic liquids (ILs) is evolving as an effective approach for pretreating intractable LB. Ionic liquids are organic salts with thermal stability. They have a substantial application like 'green solvents' and can potentially dissolve polar and non-polar organic, inorganic, and polymeric compounds. The active dissolution of cellulose and hemicellulose sugars becomes feasible when biomass is pretreated with ILs.

There are many plus points associated with the ILs. They are recyclable, chemically stable, non-flammable, non-volatile, and thermally stable (up to 400°C). Ionic liquid can dissolve solutes of fluctuating polarity. They are also involved in the production of novel chemicals and materials from biomass (Yoo *et al.*, 2017). Although IL-mediated pretreatment is an eye-catching technology for pretreating LB from the past few years, its large-scale employment has yet to be evaluated. The price of ILs is the chief barrier that can be curtailed by in-house synthesis of cheap ILs along with their recovery for reprocess. However, the efficiency of ILs gradually declines on their subsequent reuse (Liu *et al.*, 2017). Ionic liquids can either be aprotic or protic. Examples of aprotic ILs are N, N-dimethylformamide (DMF), dimethylsulphoxide (DMSO), N, N-dimethylacetamide (DMA), 1,3-dimethyl-2-imidazolidinone (DMI) and 1-butyl3-methylimidazolium chloride [bmim] [Cl]). Examples of protic ionic liquids are pyridinium formate ([Py][For]), pyridinium acetate ([Py][Ac]), pyridinium propionate ([Py][Pro]), and triethylammonium hydrogen sulfate [TEA][HSO_4]).

Aprotic ILs are now a days replaced by protic ILs, as the latter is cost effective and easier to produce. Table 2.1 depicts various ionic liquids employed for pretreating for biomass.

There are disadvantages associated with the use of ILs. The complete dissolution of LB takes place at higher temperature (more than 100°C) with prolonged processing time. This disadvantage can be crossed easily by associating ultrasound with the ILs process (Zhang *et al.*, 2015 a, b). *Fusarium oxysporum* BN is an ionic liquid-tolerant fungi that has the potential to produce ionic liquid-stable cellulase. *F. oxysporum* BN can directly convert IL pretreated rice straw to bioethanol yielding 0.125 g ethanol g^{-1} of rice straw (Xu *et al.*, 2015).

Table 2.1 Employment of various ionic liquids for biomass pretreatment.

Ionic liquid	Biomass pretreated	Recovery	Reference
Triethylammonium hydrogen sulfate [TEA] [HSO_4],	Miscanthus x giganteus	Post enzymatic hydrolysis, 85% of lignin and 100% of hemicellulose were solubilized along with 77% glucose recovered.	Brandt–Talbot *et al.* (2017)
1-H-3-imidazolium chloride [HMIM Cl]	Corn stover	Enzymatic hydrolysis of the cellulose fraction, 92.5% of conversion yield of cellulose in glucose.	Zhang *et al.* (2015a, b)
Imidazolium-based ([(C3N2)Xn]+) ionic liquid (1-butyl-3-methylimidazolium chloride [bmim][Cl])	Oil palm empty fruit bunch		Mohtar *et al.* (2017)

2.2.1.3 *Physicochemical Pretreatment Methods*

2.2.1.3.1 Steam Explosion

Steam explosion pretreatment is a physicochemical approach for the hydrolysis of the assembly of lignocellulose biomass (McMillan, 1994). Steam explosion is a mild, chemical free, cost-effective approach to pretreat LB. Water is involved in heating the biomass under pressure followed by a sudden decompression of the reaction vessel (Stelte, 2013). Temperature involved in this process varies from (160–260°C) whereas, pressure varies from 0.69–4.83 MPa. The disruption of lignocelluloses during steam explosion opens the fibers thereby leaving sugar polymers, which are vulnerable to the enzymatic attack. Hemicelluloses degradation and transformation of lignin are also the outcomes of steam explosion (Ballesteros *et al.*, 2006). Lower temperature and extended residence time favor this process. Absence of chemicals provides protection to the reaction vessel from getting corroded thereby curtailing the cost input to involve corrosion free metallurgy in this process. Sugar degradation products (furfurals) that inhibit the saccharification or fermentation are not generated. Residence time, temperature, size of chips, and moisture content are some of the crucial aspects that affect steam explosion pretreatment (Wright, 1998). As compared to other harsh pretreatment approaches, steam explosion is less effective. Enzymes are required in enormous amounts during the saccharification process. Softwood and hardwood are less effectively pretreated by this approach as they have high lignin content (Yang and Wyman, 2008).

2.2.1.3.2 Liquid Hot Water

Liquid hot water pretreatment method employs water under higher temperatures and pressures to retain its liquid state. This approach provokes the breakdown of the lignocellulose matrix along with separation. In this method, the temperature varies from 160 to 240 °C and the process gets completed in few minutes to an hour. The generated quality of sugar is governed by the temperature employed, whereas the quantity of sugar production is controlled by the time duration (Yu *et al.*, 2010). This approach of pretreating biomass depicts advantages pertaining to price standpoint, as there is no involvement of acid catalysts.

2.2.1.3.3 Ammonia Fiber Explosion

The ammonia fiber/freeze explosion (AFEX) method is a physicochemical process that allows substantial delignification of biomass along with negligible degradation of sugars. However, woody biomass residues are less effectively pretreated (Yang and Wyman, 2008). The method is a cost-effective, less energy intensive physicochemical process of pretreating biomass at moderate temperatures (60°C–100°C). Ammonia lignin's effect is lessened on enzymatic hydrolysis. Lower moisture content and moderate conditions make AFEX a beneficial process, which also supports less production of sugar degradation products. Moreover, solid material is totally recovered (Mosier *et al.*, 2005).

This process is analogous to the steam explosion pretreatment method, in which liquid anhydrous ammonia is employed at elevated pressures and moderate temperatures following rapid depressurization (Chundawat *et al.*, 2007). The need for moderate temperature makes this pretreatment a cost-effective approach as it curtails the requirement of energy input, but it simultaneously makes this process less efficient and significant when compared to the steam explosion process. Moreover, recycle and treatment of chemicals are cost intensive processes. These are some demerits associated with the process.

2.2.1.3.4 Ammonia Recycle Percolation

In the ammonia recycle percolation (ARP) strategy, 5%–15% (wt%) of aqueous ammonia is permissible to percolate over a packed bed reactor retaining the LB at a rate of about 5 ml/min. It is an improved approach compared to AFEX, as it has potential to remove 75%–85% of lignin content and to dissolve 50%–60% of the hemicellulosic material (Kim and Lee, 2005). The strategy can retain cellulosic material. In this process, the ammonia can be recycled. Herbaceous biomass (corn stover and switchgrass) has been most treated with ARP. Pretreatment of corn stover and switchgrass with ARP results in the removal of 60%–80% and 65%–85% lignin content, respectively (Iyer *et al.*, 1996). Ammonia recycle percolation (ARP) can be associated with the AFEX pretreatment process, a process that has been explored by many researchers (Kim and Lee, 2005). Release of hemicelluloses from carbohydrate–lignin complex was reported on dilute ammonia pretreatment of corn stover (Jonathan *et al.*, 2017).

2.2.1.3.5 Hydrodynamic Cavitation

The approach of hydrodynamic cavitation (HC) helps in accelerating chemical reactions (Gogate, 2016). This approach has been explored to pretreat variety of LB like sugarcane bagasse, corn stover, and reed with good outcomes (Terán–Hilares *et al.*, 2017a). The efficient removal of the lignin along with enhancement of the pore size of LB is the outcome of this pretreatment method that leads to an effective saccharification of carbohydrate fraction during its enzyme digestion. It is also a less energy-intensive process that requires fewer chemical catalysts (Gogate and Pandit, 2000). The processing time of HC is also short, which makes it an attractive option for scaling up. The system involved in HC is simple and can be modified in the continuous process (Terán–Hilares *et al.*, 2017b). When microbubbles grow and collapse, it gives rise to the cavitation phenomenon. Microbubbles are produced when fluid pressure declines below the solution vapor pressure, tailed by a fluid pressure retrieval (Gogate and Pandit, 2005). This phenomenon occurs when fluid moves through a compression device such as a venture tube or an orifice plate (Patil *et al.*, 2016). Inside the bubble exists drastic conditions of very high temperatures (10000 K) and pressures (1000 atm). Hot spots are generated in localized points when a high amount of energy is released by bubbles (Saharan *et al.*, 2013). Along with these reactions, water molecules are detached in the cavities, ensuing in the formation of potential oxidative radicals. Potential oxidative radicals (OH·, H·, HOO·, and O_2·) also get liberated in the medium when bubbles collapse (Badve *et al.*, 2014). The oxidation and degradation of lignin takes places because of these oxidative radicals. This process results in efficient elimination of lignin fraction, thereby supporting substantial enzymatic yields. The mechanical effect of HC is responsible for this higher yield. Porosity enhances during HC as an outcome of generation of shockwaves and microjets (with high velocities, as 100 m/s) when the bubble collapses. This phenomenon degrades the organic molecules, creating small holes on the biomass surface. As a result, the porosity of pretreated material increases that subsequently enhances the availability of enzyme to the cellulosic and hemicellulosic fraction (Terán–Hilares *et al.*, 2017a). Density and viscosity of fluid, process temperature, chemical catalyst (alkaline, oxidant agents, etc.), particle size, and the system conformation (number and diameter of orifices or throat size in tube) are some properties that strongly influence the cavitation phenomenon and effectiveness of the method (Terán–Hilares *et al.*, 2017b).

2.2.1.3.6 Supercritical Fluid Pretreatment

Supercritical fluid (SF)-based pretreatments are striking substitutes due to their cost effective and eco-friendly performance (Daza Serna *et al.*, 2016). This approach presents a real feasibility for getting applied at large-scale operations. Though this technology has been assessed for some decades, the last few years have witnessed it as a promising and green technology for pretreating different LB (Liang *et al.*, 2017).

Carbon dioxide is a low-cost, non-toxic recyclable that is mostly used as a supercritical fluid. The critical temperature and pressure of CO_2 is low viz. 30.95°C and 1070.37 psi, respectively. The disruption of the crystalline structure of biomass occurs concurrently with an effective lignin elimination when the pretreatment process employs supercritical CO_2 ($scCO_2$). This approach ameliorates the conversion of cellulose (Liang *et al.*, 2017). This approach is an eco-friendly one as the usage of CO_2 curtails the impact of greenhouse on the atmosphere. Moisture content of biomass plays a crucial role in this pretreatment approach. When CO_2 combines with water, carbonic acid is produced, which may lead to hydrolysis of hemicellulose. The moisture content of a rice husk was adjusted at 75% employing a water:ethanol mixture as a co-solvent when it was exposed to the $scCO_2$ pretreatment. The pretreatment took place at 80°C and 270 bars (Daza Serna *et al.*, 2016). In these parameters of pretreatment, 90.6% of lignin removal and 55% of enzymatic hydrolysis was attained.

2.2.1.4 *Biological Pretreatment*

There are many microorganisms that have the potential to degrade cellulose, hemicellulose, and lignin into their constitutive sugars. Prominent among them are bacteria, actinomycetes, and fungi. This process is operated in mild conditions of temperatures (Chandel *et al.*, 2013). These methods are cost-effective and eco-friendly approaches in obtaining the goals of pretreatment, thereby circumventing various challenges faced by other chemical approaches of pretreatment. Various constituents of biomass like lignin, cellulose, hemicellulose, and polyphenols get hydrolyzed/degraded during biological pretreatment (Sindhu *et al.*, 2016). Glucose, arabinose, and xylose are mono sugars that get generated because of bacterial and fungal hydrolysis of cellulose and hemicellulose. Prominent fungi involved in this task are brown, white, and soft-rot fungi with an efficiency to degrade lignin and hemicellulose. White-rot fungi is potentially superior than other fungi involved. Biological pretreatment needs no acids, alkali, or any reactive species.

Monomeric sugars like glucose, arabinose, and xylose are liberated when cellulose and hemicellulose get hydrolyzed by the action of various species of bacteria and fungi. Selective degradation of lignin and hemicelluloses occurs by the action of brown, white, and soft-rot fungi, among which white-rot fungi is highly efficient. Biological pretreatments occur in ambient conditions of temperature and pressure and is thus different from methods of physical and chemical pretreatment. Moreover, biological pretreatments require no acid, alkali, or reactive species. The main plus point of biological methods is that this method is less energy intensive and can be performed at mild conditions (Akhtar *et al.*, 2016). The only demerit associated with biological pretreatment process is the low rate of hydrolysis. Microorganisms have potential to degrade lignin, hemicellulose, cellulose and carbohydrates by the virtue of the enzymes (lignocellulosic enzymes) secreted by them, and hence it is needed to explore such microorganisms with a potential to degrade cell wall (Kubicek *et al.*, 2014). Table 2.2 illustrates various microorganisms and their enzymes that are involved in biological pretreatment.

Many researchers have explored a variety of microorganisms from different habitats for their potential to explore lignocellulosic enzymes. For example, the *Aspergillus sydowii* strain MS-19 was isolated from Antarctic region showing substantial activity of cell wall degradation (Cong *et al.*, 2017). *Trichoderma reesei* (Saloheimo *et al.*, 2002), *Neurospora crassa* (Tian *et al.*, 2009) and *Rhizopus oryzae* (Lara–Marquez *et al.*, 2011) are some fungi that were successfully exploited. Important bacteria associated with biological pretreatment are *Pseudomonas aeruginosa, P. stutzeri, Streptomyces badius, S. setonii, Rhodococcus ruber, Comamonas acidovorans, Clostridium thermocellum, Butyrivibrio fibrisolvens*). Table 2.3 illustrates various examples of microorganisms that can degrade different fractions of LB.

Table 2.2 Microorganisms and their enzymes involved in biological pretreatment.

Microorganism	Plant attacked	Lignocellulosic component degrades	Microbial enzymes	Reference
Bacillus subtilis strain 168		Pectin and polygalacturonan		Ochiai *et al.* (2007)
Pectobacterium, Klebsiella, Serratia, Enterobacter, Citrobacter, Providencia, and *Pseudomonas*	Soft rot of orchid		Pectate lyase, polygalacturonase, cellulase, and protease	Joko *et al.* (2014)
Macrophomina phaseolina			Pectinases, cellulases, hemicellulases, and laccase	Ramos *et al.* (2016)

(Continued)

Table 2.2 Microorganisms and their enzymes involved in biological pretreatment. (*Continued*)

Microorganism	Plant attacked	Lignocellulosic component degrades	Microbial enzymes	Reference
Alternaria, i.e., *A. alternata*, *A. macrospora* MKP2, *A. macrospora* MKP4, *Alternaria sp.* PMK1, and *Alternaria sp.* PMK2	Leaves of Parthenium plant	Degradation of polymers available in the host plants' cell wall	Cellulase, laccase and lignin peroxidase	Kaur and Aggarwal (2017)
Phanerochaete chrysosporium, *Ceriporia lacerata*, *Cyathus stercolerus*, *Ceriporiopsis subvermispora*, *Pycnoporus cinnarbarinus*, *Pleurotus ostreaus* and *P. chrysosporium*			Lignin-degrading enzymes i.e., lignin peroxidases and manganese-dependent peroxidases	Chandel *et al.* (2013)
Phlebiaradiata, *P. floridensis* and *Daedalea flavida*	Wheat straw	Hydrolyze lignin		Yang and Wyman (2008)
Ceriporiopsis subvermispora	Various wood species	Hydrolyze lignin	Manganese peroxide, as well as laccase	Yang and Wyman (2008)

Table 2.3 Examples of various microorganisms degrading different fractions of LB.

Lignin-degrading fungi	*Phanerochaete chrysosporium, Pycnoporus cinnabarinus, Phlebia spp., Echinodontium taxodii, Irpexlacteus* and *Pycnoporus sanguineus*
Carbohydrate-degrading fungi	*Ceriporiopsis subvermispora, Phlebiabrevispora, P. floridensis, P. radiata, Echinodontium taxodii, Gonoderma sp., Oxysporus sp., Trametes versicolor, Pleurotus sajor-caju* and *Trichoderma reesei*
Cellulosic material-degrading fungi	*Serpula lacrymans, Coniophora puteana, Meruli poria incrassata, Laetoporus sulphureus,* and *Gleophyllum trabeum*
Lignocellulosic biomasses-degrading bacteria	*Bacillus circulans* and *Sphingomonas paucimobilis, Cellulomonas, Zymomonas spp., Bacillus macerans, Cellulomonas cartae,* and *C. uda*

References

Akhtar, N., Gupta, K., Goyal, D., Goyal, A., Recent advances in pretreatment technologies for efficient hydrolysis of lignocellulosic biomass. *Environ. Prog. Sustain. Energy*, 35, 2, 489–511, 2016.

Antunes, F.A.F., Chandel, A.K., Terán Hilares, R., Ingle, A.P., Rai, M., Milessi, T.S., da Silva, S.S., dos Santos, J.C., Overcoming challenges in lignocellulosic biomass pretreatment for second generation (2G) sugar production: emerging role of nano, biotechnological and promising approaches. *3 Biotech*, 9, 230, 2019, https://doi.org/10.1007/s13205-019-1761-1.

Aswathy, U.S., Sukumaran, R.K., Lalitha, D., Rajeshree, K.P., Singhania, R.R., Pandey, A., Bio-ethanol from water hyacinth biomass: an evaluation of enzymatic saccharification strategy. *Bioresour. Technol.*, 101, 925–930, 2010.

Auxenfans, T., Crônier, D., Chabbert, B., Paës, G., Understanding the structural and chemical changes of plant biomass-following steam explosion pretreatment. *Biotechnol. Biofuels*, 10, 36, 2017. https ://doi.org/10.1186/s13068-017-0718-z.

Badve, M.P., Gogate, P.R., Pandit, A.B., Levente, C., Hydrodynamic cavitation as a novel approach for delignification of wheat straw for paper manufacturing. *Ultrason. Sonochem.*, 21, 162–168, 2014.

Ballesteros, I., Negro, M.A.J., Oliva, J.M., Cabanas, A., Manzanares, P., Ballesteros, M., Ethanol production from steam-explosion pretreated wheat straw. *Appl. Biochem. Biotechnol.*, 130, 1–3, 496–508, 2006.

Bernardinelli, O.D., Lima, M.A., Rezende, C.A., Polikarpov, I., Azevedo, E.R., Quantitative 13C Multi CP solid-state NMR as a tool for evaluation of cellulose crystallinity index measured directly inside sugarcane biomass. *Biotechnol. Biofuels*, 8, 110, 2015.

Bhatia, L. and Johri, S., Fourier transform infrared mapping of peels of *Litchi chinensis* after acid treatment and its SSF for ethanol production by *Pachysolen tannophilus* MTCC 1077- An economic & ecological venture. *Indian J. Biotechnol.*, 16, 444–456, 2017.

Bjerre, A.B., Olesen, A.B., Fernqvist, T., Pretreatment of wheat straw using combined wet oxi- dation and alkaline hydrolysis resulting in convertible cellulose and hemicellulose. *Biotechnol. Bioeng.*, 49, 568–577, 1996.

Brandt-Talbot, A., Gschwend, F.J.V., Fennell, P.S., Lammens, T.M., Tan, B., Weale, J., An economically viable ionic liquid for the fractionation of lignocellulosic biomass. *Green Chem.*, 19, 1–26, 2017.

Cadoche, L. and Lopez, G.D., Assessment of size reduction as a preliminary step in the production of ethanol from lignocellulosic wastes. *Biol. Wastes*, 30, 153–157, 1989.

Chandel, A.K., Antunes, F.A.F., Anjos, V., Bell, M.J.V., Rodrigues, L.N., Polikarpov, I., Multi-scale structural and chemical analysis of sugarcane bagasse in the process of sequential acid-base pretreatment and ethanol production by Scheffersomyces shehatae and Saccharomyces cerevisiae. *Biotechnol. Biofuels*, 7, 63, 2014.

Chandel, A.K., Kapoor, R.K., Singh, A., Kuhad, R.C., Detoxification of sugarcane bagasse hydrolysate improves ethanol production by *Candida shehatae* NCIM 3501. *Bioresour. Technol.*, 98, 1947–1950, 2007.

Chandel, A.K., Silva, S.S., Singh, O.V., Detoxification of lignocellulose hydrolysates: Biochemical and metabolic engineering toward white biotechnology. *Bioenergy Res.*, 6, 388–401, 2013.

Cheah, W.Y., Show, P.L., Juan, J.C., Chang, J.S., Ling, T.C., Waste to energy: the effects of Pseudomonas sp. on Chlorella sorokiniana biomass and lipid productions in palm oil mill effluent. *Clean Technol. Environ. Policy*, 20, 2037–2045 2018, https://doi.org/10.1007/s10098-018-1505-7.

Chundawat, S.P.S., Venkatesh, B., Dale, B.E., Effect of particle size based separation of milled corn stover on AFEX pretreatment and enzymatic digestibility. *Biotechnol. Bioeng.*, 96, 2, 219, 2007.

Cong, B., Wang, N., Liu, S., Liu, F., Yin, X., Shen, J., Isolation, characterization and transcriptome analysis of a novel antarctic Aspergillus sydowii strain MS-19 as a potential lignocellulosic enzyme source. *BMC Microbiol.*, 17, 129, 2017.

Cortes-Tolalpa, L., Jiménez, D.J., Brossi L., Salles, J.F., Elsas, J.D., Different inocula produce distinctive microbial consortia with similar lignocellulose degradation capacity. *Appl. Microbiol. Biotechnol.*, 100, 7713–7725, 2016. DOI 10.1007/s00253-016-7516-6.

Cucchiella, F., D'Adamo, I., Gastaldi, M., Financial analysis for investment and policy decisions in the renewable energy sector. *Clean Technol. Environ. Policy*, 17, 4, 887–904, 2014.

Dawson, L. and Boopathy, R., Use of post-harvest sugarcane residue for ethanol production. *Biores. Technol.*, 98, 1695–1699, 2007.

Daza Serna, L.V., OrregoAlzate, C.E., Cardona Alzate, C.A., Supercritical fluids as a green technology for the pretreatment of lignocellulosic biomass. *Bioresour. Technol.*, 199, 113–120, 2016.

De Moraes-Rocha, G.J., Martin, C., Soares, I.B., Maior, A.M.S., Baudel, H.M., De Abreu, C.A.M., Dilute mixed-acid pretreatment of sugarcane bagasse for ethanol production. *Biomass Bioenergy*, 35, 663–670, 2010.

Driemeier, C., Pimenta, M.T.B., Rocha, G.J.M., Oliveira, M.M., Mello, D.B., Maziero, P., Evolution of cellulose crystals during prehydrolysis and soda delignification of sugarcane lignocellulose. *Cellulose*, 18, 1509–1519, 2011.

Ebringerova, A. and Hromadkova, Z., Effect of ultrasound on the extractability of cornbran hemi- celluloses. *Ultrason. Sonochem.*, 9, 4, 225–229, 2002.

Elmore, J.R., Furches, A., Wol, G.N., Gorday, K., Guss, A.M., Development of a high efficiency integration system and promoter library for rapid modification of Pseudomonas putida KT2440. *Metab. Eng. Commun.*, 5, 1–8, 2017.

Fonseca, B.G., Moutta, R.O., Ferraz F.O., Vieira E.R., Nogueira A.S., Baratella B.F., Rodrigues L.C., Hou-Rui, Z., Silva, S.S., Biological detoxification of different hemicellulosic hydrolysates using *Issatchenkia occidentalis* CCTCC M 206097 yeast. *J. Ind. Microbiol. Biotechnol.*, 38, 199–207, 2011.

Goel, A. and Wati, L., Ethanol production from Rice (*Oryza sativa*) straw by simultaneous saccharification and cofermentation. *Indian J. Exp. Biol.*, 54, 525–529, 2016.

Gogate, P.R., Greener processing routes for reactions and separations based on use of ultrasound and hydrodynamic cavitation, in: *Alternative energy sources for green chemistry*, G. Stefanidis and A. Stankiewicz (Eds.), Thomas Graham House, RSC Publishing Blog, Cambridge, 2016.

Gogate, P.R. and Pandit, A.B., Engineering design methods for cavitation reactors II: hydrodynamic cavitation. *AlChE J.*, 46, 1641–1649, 2000.

Gogate, P.R. and Pandit, A.B., A review and assessment of hydrodynamic cavitation as a technology for the future. *Ultrason. Sonochem.*, 12, 21–27, 2005.

Goldstein, I. and Easter, J., An improved process for converting cellulose to ethanol. *Tappi J.*, 28, 135, 1992.

Ibrahim, M.M., El-Zawawy, W.K., Abdel-Fattah, Y.R., Soliman, N.A., Agblevor, F.A., Comparison of alkaline pulping with steam explosion for glucose production from rice straw. *Carbohydr. Polym.*, 83, 2, 720–726, 2011.

Imai, M., Ikari, K., Suzuki, I., High-performance hydrolysis of cellulose using mixed cellulase species and ultrasonication pretreatment. *Biochem. Eng. J.*, 17, 79–83, 2004.

Iyer, P.V., Wu, Z.W., Kim, S.B., Lee, Y.Y., Ammonia recycled percolation process for pretreatment of herbaceous biomass. *Appl. Biochem. Biotechnol.*, 57–58, 121–132, 1996.

Joko, T., Subandi, A., Kusumandari, N., Wibowo, A., Priyatmojo, A., Activities of plant cell wall-degrading enzymes by bacterial soft rot of orchid. *Arch. Phytopathol. Plant Protect.*, 47, 10, 1239–1250, 2014.

Jönsson, L.J. and Martín, C., Pretreatment of lignocellulose: Formation of inhibitory by-products and strategies for minimizing their effects. *Bioresour. Technol.*, 199, 103–112, 2016.

Jonathan, M.C., DeMartini, J., Van StigtThans, S., Hommes, R., Kabel, M.A., Characterisation of non-degraded oligosaccharides in enzymatically hydrolysed and fermented, dilute ammonia-pretreated corn stover for ethanol production. *Biotechnol. Biofuels*, 10, 112, 2017, DOI: 10.1186/s13068-017-0803-3.

Kaur, M. and Aggarwal, N.K., Screening of Alternaria pathogens associated with Parthenium hysterophorus for the production of lignocellulolytic enzymes. *Bioeng. Biosci.*, 5, 1, 14–23, 2017.

Kim, T.H. and Lee, Y.Y., Pretreatment and fractionation of cornstover by ammonia recycle percolation process. *Bioresour. Technol.*, 96, 18, 2007–2013, 2005.

Kubicek, C.P., Starr, T.L., Glass, N.L., Plant cell wall-degrading enzymes and their secretion in plant-pathogenic fungi. *Annu. Rev. Phytopathol.*, 52, 427–451, 2014.

Lara-Marquez, A., Zavala-Paramo, M.G., Lopez-Romero, E., Camacho, H.C., Biotechnological potential of pectinolytic complexes of fungi. *Biotechnol. Lett*, 33, 859–868, 2011.

Liang, J., Chen, X., Wanga, L., Weia, X., Wang, H., Lu, S., Subcritical carbon dioxide-water hydrolysis of sugarcane bagasse pith for reducing sugars production. *Bioresour. Technol.*, 228, 147–155, 2017.

Liu, Y., Fredrickson, J.K., Sadler, N.C., Nandhikonda, P., Smith, R.D., Wright, A.T., Advancing understanding of microbial bioenergy conversion processes by activity-based protein profiling. *Biotechnol. Biofuels*, 8, 156, 2015, DOI: 10.1186/s13068-015-0343-7.

Liu, Z., Longfei, L., Cheng, L., Airong, X., Saccharification of cellulose in the ionic liquids and glucose recovery. *Renew. Energy*, 106, 99–102, 2017.

Maurya, D.P., Singla, A., Negi, S., An overview of key pretreatment processes for biological conversion of lignocellulosic biomass to bioethanol. *3 Biotech.*, 5, 597–609, 2015.

Mc Millan, J.D., Pretreatment of lignocellulosic biomass, in: *Enzymatic conversion of biomass for fuels production*, M.E. Himmel, J.O. Baker, R.P. Overend (Eds.), pp. 292–324, American Chemical Society, Washington, DC, 1994.

Millet, M.A., Baker, A.J., Scatter, L.D., Physical and chemical pretreatment for enhancing cellulose saccharification. *Biotech. Bioeng. Symp.*, 6, 125–153, 1976.

Mirmohamadsadegh, S., Chen, Z., Wan, C., Reducing biomass recalcitrance via mild sodium carbonate pretreatment. *Bioresour. Technol.*, 209, 386–390, 2016.

Mohtar, S.S., Busu, TNZTM, Noor, A.M.M., Shaari, N., Mat, H., An ionic liquid treatment and fractionation of cellulose, hemicellulose and lignin from oil palm empty fruit bunch. *Carbohydr. Polym.*, 166, 291–299, 2017.

Mosier, N., Wyman, C.E., Dale, B.D., Elander, R.T., Lee, Y.Y., Holtzapple, M., Ladisch, C.M., Features of promising technologies for pretreatment of lignocellulosic biomass. *Bioresour. Technol.*, 96, 673–686, 2005.

Ochiai, A., Itoh, T., Kawamata, A., Hashimoto, W., Murata, K., Plant cell wall degradation by saprophytic Bacillus subtilis strains: Gene clusters responsible for rhamnogalacturonan depolymerization. *Appl. Environ. Microbiol.*, 73, 12, 3803–3813, 2007.

Onoghwarite, O.E., Obiora, N.V., Ben, E.A., Moses, N.O.E., Bioethanol production from corn stover using Saccharomyces cerevisiae. *Int. J. Sci. Eng. Res.*, 7, 8, 2016.

Pappas, C., Tarantilis, P.A., Daliani, I., Mavromoustakos, T., Polissiou, M., Comparison of classical and ultrasound-assisted isolation procedures of cellulose from kenaf (Hibiscus cannabinus L.) and eucalyptus (Eucalyptus rodustrus Sm.). *Ultrason. Sonochem.*, 9, 1, 19–23, 2002.

Parawira, W. and Tekere, M., Biotechnological strategies to overcome inhibitors in lignocellulose hydrolysates for ethanol production: review. *Crit. Rev. Biotechnol.*, 31, 20–31, 2011.

Pathak, B.S. and Chandel, A.K., Feedstock Transportation, Agricultural Processing, Logistic from Farm to Bio-refinery: Recent Developments, Mechanization, and Cost Analysis, in: *Sustainable Biofuels Development in India*, A.K. Chandel and R.K. Sukumaran (Eds.), Springer, Cham, 2017, DOI: 10.1007/978-3-319-50219-9_9.

Patil, P.N., Gogate, P.R., Csoka, L., Dregelyi-Kiss, A., Horvath, M., Intensification of biogas production using pretreatment based on hydrodynamic cavitation. *Ultrason. Sonochem.*, 30, 79–86, 2016.

Philippini, R.R., Martiniano, S.E., Chandel, A.K., de Carvalho, W., Silva, S.S., Pretreatment of Sugarcane Bagasse from Cane Hybrids: Effects on Chemical Composition and 2G Sugars Recovery, in: *Waste and Biomass Valorization*, 2017, https://doi.org/10.1007/s12649-017-0162-0.

Ramos, A.M., Gally, M., Szapiroa, G., Itzcovicha, T., Carabajal, M., Levina, L., *In vitro* growth and cell wall degrading enzyme production by Argentinean isolates of Macrophomina phaseolina, the causative agent of charcoal rot in corn. *Rev. Argent. Microbiol.*, 48, 4, 267–273, 2016.

Reginatto, V. and Antônio, R.V., Fermentative hydrogen production from agroindustrial lignocellulosic substrates. *Braz. J. Microbiol.*, 46, 2, 323–335, 2015. DOI: http://dx.doi.org/10.1590/S1517-838246220140111.

Rolz, C., Ultrasound effect of enzymatic saccharification. *Biotech Lett.*, 8, 2, 131–136, 1986.

Ruffell, J., Levie, B., Helle, S., Duff, S., Pretreatment and enzymatic hydrolysis of recovered fibre for ethanol production. *Bioresour. Technol.*, 101, 7, 2267–2272, 2010.

Saharan, V.K., Manava, R., Aqeel, M., Pandit, A.B., Effect of geometry of hydrodynamically cavitating device on degradation of orange-G. *Ultrason. Sonochem.*, 20, 345–353, 2013.

Saloheimo, M., Paloheimo, M., Hakola, S., Pere, J., Swanson, B., Nyyssönen, E., Swollenin, a Trichoderma reesei protein with sequence similarity to the plant expansins, exhibits disruption activity on cellulosic materials. *Eur. J. Biochem.*, 269, 4202–4211, 2002.

Schmidt, A.S., Mallon, S., Thomsen, A.B., Hvilsted, S., Lawther, J.M., Comparison of the chemical properties of wheat straw and beech fibers following alkaline wet oxidation and laccase treatments. *J. Wood Chem. Technol.*, 22, 1, 39–53, 2002.

Scott, W.E. and Gerber, P., Using ultrasound to deink xerographic waste. *Tappi J.*, 78, 125–130, 1995.

Shafizadeh, F. and Bradbury, A.G.W., Thermal degradation of cellulose in air and nitrogen at low temperatures. *J. Appl. Poly. Sci.*, 23, 1431–1442, 1979.

Sills, D.L. and Gossett, J.M., Assessment of commercial hemicellulases for saccharification of alkaline pretreated perennial biomass. *Bioresour. Technol.*, 102, 2, 1389–1398, 2011.

Sindhu, R., Binod, P., Pandey, A., Biological pretreatment of lignocellulosic biomass: an overview. *Bioresour. Technol.*, 199, 76–82, 2016.

Stelte, W., *Steam Explosion for Biomass Pre-Treatment*, Danish Technological Institute, Denmark, 2013.

Sun, R.C. and Tomkinson, J., Comparative study of lignin isolated by alkali and ultrasound-assisted alkali extractions from wheat straw. *Ultrason. Sonochem.*, 9, 2, 85–93, 2002.

Taherzadeh, M.J. and Karimi, K., Pretreatment of lignocellulosic wastes to improve ethanol and biogas production: a review. *Int. J. Mol. Sci.*, 9, 1621–1651, 2008.

Terán-Hilares, R., de Almeida, G.F., Ahmed, MA., Antunes, F.A.F., da Silva, S.S., Han, J., *et al.*, Hydrodynamic cavitation as an efficient pretreatment method for lignocellulosic biomass: A parametric study. *Bioresour. Technol.*, 235, 301–308, 2017a.

Terán-Hilares, R., Ramos, L., Silva, S.S., Dragone, G., Mussatto, S.I., Santos, J.C., Hydrodynamic cavitation as a strategy to enhance the efficiency of lignocellulosic biomass pretreatment. *Crit. Rev. Biotechnol.*, 38, 483–493, 2017b.

Tian, C., Beeson, W.T., Iavarone, A.T., Sun, J., Marletta, M.A., Cate, J.H.D., Systems analysis of plant cell wall degradation by the model filamentous fungus Neurospora crassa. *Proc. Natl. Acad. Sci.*, 106, 22157–22162, 2009.

Valdivia, M., Galan, J.L., Laffarga, J., Ramos, J.L., Biofuels 2020: Biorefineries based on lignocellulosic materials. *Microb. Biotechnol.*, 9, 5, 585–594, 2016. doi:10.1111/1751-7915.12387.

Vinatoru, M., Toma, M., Mason, T.J., Ultrasonically assisted extraction of bioactive principles from plants and their constituents, in: *Advances in sonochemistry*, 5th, T.J. Mason (Ed.), pp. 209–248, JAI Press, London, 1999.

Waghmare, P.R., Khandare, R.V., Jeon, B.H., Govindwar, S.P., Enzymatic hydrolysis of biologically pretreated sorghum husk for bioethanol production. *Biofuel Res. J.*, 19, 846–853, 2018.

Wang, V.B., Chua, S.L., Cai, Z., A stable synergistic microbial consortium for simultaneous azo dye removal and bioelectricity generation. *Bioresour. Technol.*, 155, 71–76, 2014.

Wood, B.E., Aldrich, H.C., Ingram, L.O., Ultrasound stimulates ethanol production during the simultaneous saccharification and fermentation of mixed waste office paper. *Biotechnol. Prog.*, 13, 3, 232–237, 1997.

Wright, J.D., Ethanol from biomass by enzymatic hydrolysis. *Chem. Eng. Prog.*, 84, 8, 62–74, 1998.

Xu, C., Singh, D., Dorgan, K.M., Zhang, X., Chen, S., Screening of ligninolytic fungi for biological pretreatment of lignocellulosic biomass. *Can. J. Microbiol.*, 61, 10, 745–752, 2015.

Yang, B. and Wyman, C., Pretreatment: the key to unlocking low-cost cellulosic ethanol. *Biofuels Bioprod. Bioref.*, 2, 26–40, 2008.

Yoo, C.G., Pu, Y., Ragauskas, A.J., Ionic liquids: Promising green solvents for lignocellulosic biomass utilization. *Curr. Opin. Green Sustain. Chem.*, 5, 5–11, 2017.

Yu, G., Yano, S., Inoue, H., Inoue, S., Endo, T., Sawayama, S., Pretreatment of rice straw by a hot-compressed water process for enzymatic hydrolysis. *Appl. Biochem. Biotechnol.*, 160, 2, 539–551, 2010.

Yu, J. and Zhaoxian, X., Process integration for ethanol production from corn and corn stover as mixed substrates. *Bioresour. Technol.*, 279, 2019, DOI: 10.1016/j.biortech.2019.01.112.

Zhang, C., Xu, W., Yan, P., Liu, X., Zhang, C., Overcome the recalcitrance of eucalyptus bark to enzymatic hydrolysis by concerted ionic liquid pretreatment. *Process Biochem.*, 50, 2208–2214, 2015a.

Zhang, W., Qiu, J., Feng, H., Zang, L., Sakai, E., Increase in stability of cellulase immobilized on functionalized magnetic nanospheres. *J. Magn. Mater.*, 375, 117–123, 2015b.

Zhao, Y., Wang, Y., Zhu, J.Y., Ragauskas, A., Deng, Y., Enhanced enzymatic hydrolysis of spruce by alkaline pretreatment at low temperature. *Biotechnol. Bioeng.*, 99, 6, 1320–1328, 2008.

Zhu, J.Y., Pan, X.J., Wang, G.S., Gleisner, R., Sulfite pretreatment for robust enzymatic saccharification of spruce and red pine. *Bioresour. Technol.*, 100, 2411–2418, 2009.

3

Current Advancements and Challenges for Biomethane Production

Prakash Kumar Sarangi

College of Agriculture, Central Agricultural University, Imphal, Manipur, India

Abstract
Most energy carriers in automobiles and industrial sectors are governed by fossil-based fuels. These not only create environmental problems but also face instability due to its limited supply and demographic parameters, thereby globally seeking alternative fuels in the last few decades. Hence, research has led to the proliferation of various biomasses, as alternative and sustainable material for future fuel production to mitigate fossil-based environmental threats. Like other biofuels, biogas, derived from biobased sources, is considered as an effective solution in meeting future challenges raised by fossil fuel, unusual climate change, and greenhouse gas emission. Biomethane is bioenergy gas, which can be produced by the valorization of waste biomass using methanogenesis technologies. In this chapter, we focus on bioconversion technologies for producing biomethane from waste biomass. This chapter discusses the production of biogas with emphasis on the different biomass utilization for various industrial and domestic applications. Technological developments and future challenges are also discussed inside the chapter.

***Keywords*:** Bioconversion, bioethanol, biogas, waste, anaerobic digestion

3.1 Introduction

The global energy crisis raises a significant issue due to industrial developments and urbanization, which leads to a generation of major greenhouse gases. Excessive use of conventional energy sources like fossil-based fuels create foremost environmental pollution, thereby forcing researchers and academicians to search new green technologies to minimize its excessive use

Prakash Kumar Sarangi. Waste Biomass to Alternative Fuels, (45–66) © 2025 Scrivener Publishing LLC

(Nanda *et al.*, 2020; Pasin *et al.*, 2020; Siang *et al.*, 2020; Bhatia *et al.*, 2020; Sarangi *et al.*, 2020c). Hence, biobased renewable sources can be targeted to produce energy and fuel sources. Thus, developing sustainable process technology to utilize waste biomass and not harm the environment can be focused on (Sarangi *et al.*, 2020b), which stands a big challenge for the researchers around the world (Sarangi *et al.*, 2019). Thus, waste biomass from various processing and agro-industries can be utilized to produce valuable biochemicals along with green energy for sustainable economic development (Sarangi *et al.*, 2009a; Sarangi *et al.*, 2009b; Sarangi *et al.*, 2010; Sarangi and Sahoo, 2010a, Sarangi and Sahoo, 2010b; Sarangi and Sahoo, 2010c). Most of these by-products contain three major structural polymers such as cellulose, hemicellulose, and lignin. A high proportion of this waste material is carbohydrate and phenolic in nature. Hence, biological degradation has become an increasingly popular alternative for the treatment of agricultural, industrial, organic, as well as toxic waste by researchers and academicians worldwide. Cellulose and hemicellulose being the major constituents can be referred to as valuable resources for several reasons, mainly because they can be bio-converted easily into valuable products.

Various by-products from agro-industries like corn cobs, pineapple waste, bran, and straw can be focused as a good source for bioconversion into fuels and value-added products through microbial catalytic interventions (Sarangi *et al.*, 2019; Sarangi and Nanda, 2019a; Sarangi and Nanda, 2019b; Sarangi and Nanda, 2019c). Most petroleum fuels are mixtures of hydrocarbons, which can also be produced from carbohydrates by removing the oxygen. Hence biomass, which is rich in carbohydrates, can also be effectively converted into hydrocarbons. Biomass can be transformed into biofuel such as bioethanol, biodiesel, and biogases to use it widely for shipping, transport, numerous industrial and domestic applications. Lots of developments in the production of biofuel and biogas have been reported in recent years. Biomethane or biogas has the potential to be a biofuel source for future global bioeconomy purposes due to its production from waste biomass by specific microbial communities. In rural sectors, it can be used as gobar gas for production of household electricity. In this chapter, the production and biochemistry of process technologies about biomethane are described along with future challenges and advancements.

3.2 Production Methods of Methane

As far as the energy point is concerned, methane (CH_4) plays a vital role in the mitigation of a potential energy crisis in the future generated from biological and chemical sources. Chemical production of CH_4 is also accepted as far as the industrial point is concerned. A comparison of two processes is shown in Figure 3.1. Developing an efficient and economical methanation to convert coal to synthetic natural gas is a global focus (Kopyscinski *et al.*, 2010). In the presence of hydrogen (H_2), carbon dioxide (CO_2) and carbon monoxide are converted into CH_4 in the chemical methanation process. The major attributes of the methanation process of COx are explored for use as an alternative fuel. Methanation has been considered the most important technology to produce synthetic natural gas since the 1970s. The famous methanation reaction is called Haldor Topsoe's process, where H_2 and carbon monoxide are converted into CH_4 in the ratio of 3:1 in the presence of nickel-based catalysts.

There are various methods such as gasification, hydromethanation, supercritical water gasification, and absorption enhanced reforming adapted for production of CH_4. During the gasification process, biomass used is gasified at very high temperatures and pressures to produce H_2-rich syngas, where CH_4 is often a component of the product gases. During supercritical water gasification, when water is reacted with the feedstock in its supercritical, thereby enhancing the product efficiency. Hydromethanation, on

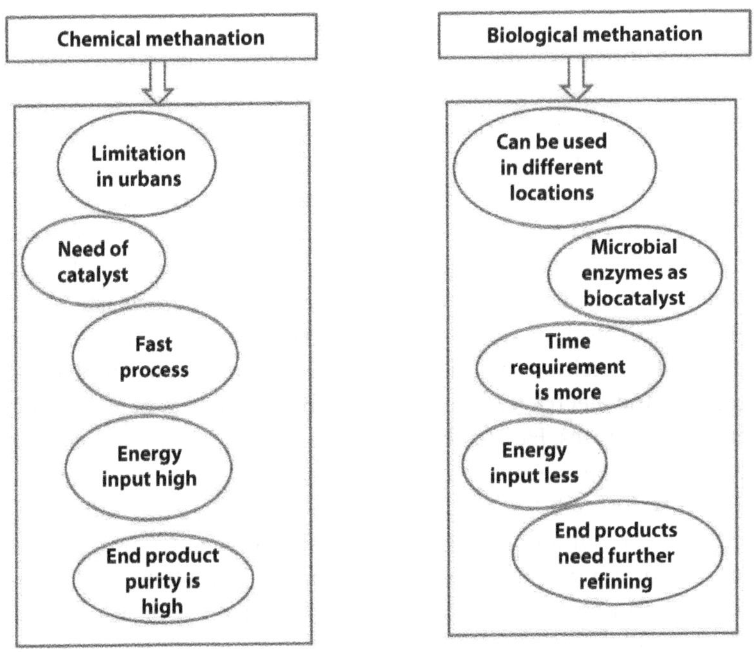

Figure 3.1 A comparison between the chemical and biological methanation process.

the other hand, takes place through an injection of pressurized steam into the fluidized mixture of catalyst and carbon-rich feedstock. Though conventional catalysts are used in the methanation process, nanoscale catalysts have been explored as effective sources in methanation. In this case, Nickel, molybdenum, ruthenium and tungsten are doped inside or on the surface of nanotubes so that these are used to enhance CH_4 yields. Through the biological process of biomethanation, biogas is produced by subjecting organic matter to microbial fermentation in the absence of oxygen, which is referred to as the 'anaerobic digestion' (AD).

Biogas production from organic waste biomass is assumed to share about 25% of the overall renewable energy budget (Holm–Nielsen *et al.*, 2009). Various waste biomass has been utilized for biogas production. These include dung, sewage sludge, several biodegradable waste from farms and households along with municipal waste, plant shreds, agro-waste, and kitchen waste. Biomethane is considered an eco-friendly alternative energy source in rural sectors to resolve energy-related limitations and challenges. The establishment of biogas plants can provide a cleaner environment, not only through energy crisis mitigation, but also with employment in rural sectors. Biomass like rice straw and pineapple waste (Sarangi and Nanda, 2019b) may be utilized to produce biogas through the biomethanation process.

Biogas acts as a renewable energy resource, which can be used as the alternative solution for the world's future energy demands by managing waste and reducing the greenhouse gas (GHG) emissions. It is considered as carbon neutral as the carbon in biogas comes from organic matter. It is the energy, which is produced from cow dung by decomposing organic materials, which occur in the absence of oxygen. This is called gobar gas as it is derived from gobar or cow dung. This can be used directly for cooking and lighting. Biogas is regarded as an eco-friendly and renewable energy source found buried in cattle dung, kitchen waste, agro-waste, residues, and municipal biodegradable waste. Biogas from a biodigester is composed of CH_4, which can be utilized as the substitute for fossil fuel towards energy, fuel, and electricity. Thereby, it can be utilized for heating, cooking, and driving purposes. Hence, this can provide an opportunity to turn plentily available, organic waste biomass into a useful energy source towards sustainability. Though the production of biogas from cattle dung and animal manure is common, still then, it can be produced from organic waste biomass including municipal waste, food processing waste, and industrial waste. Solid and semisolid waste materials from plant or animal sources can also be utilized for production of biogas.

3.3 Production of Biomethane and Biochemistry of Processes

Microbial degradation of organic waste without oxygen is known as anaerobic digestion (AD). The conversion of organic substances to CO_2 and CH_4 gases occurs beside a series of biochemical reactions during an anaerobic process. The outcome of such a reaction initiates the breakdown of organic matters during the process of digestion, and the only possibility is because of anaerobic microorganisms. Under the anaerobic process, most organic matter, which is biodegradable in nature, gets converted to gas while only a very small amount (about 10%) is converted to new cell mass by microbial growth. The production of CH_4 by AD could be an economic advantage over aerobic digestion as CH_4 gas can be used to run an effluent treatment plant. The AD process of organic materials involves the absence of oxygen by anaerobic bacteria wherein waste biomass is converted into biogas (biomethane) by four different steps such as hydrolysis, acidogenesis, acetogenesis, and methanogenesis.

The first step involves a group of microorganisms that secrete enzymes, which hydrolyzes polymeric materials to monomers such as glucose and amino acids. These are further converted by acetogenic bacteria to higher volatile fatty acids, H_2, and acetic acid. Finally, the third group of bacteria called methanogens convert H_2, CO_2, and acetate, to CH_4. An outline of the complete AD process is shown below in Figure 3.2.

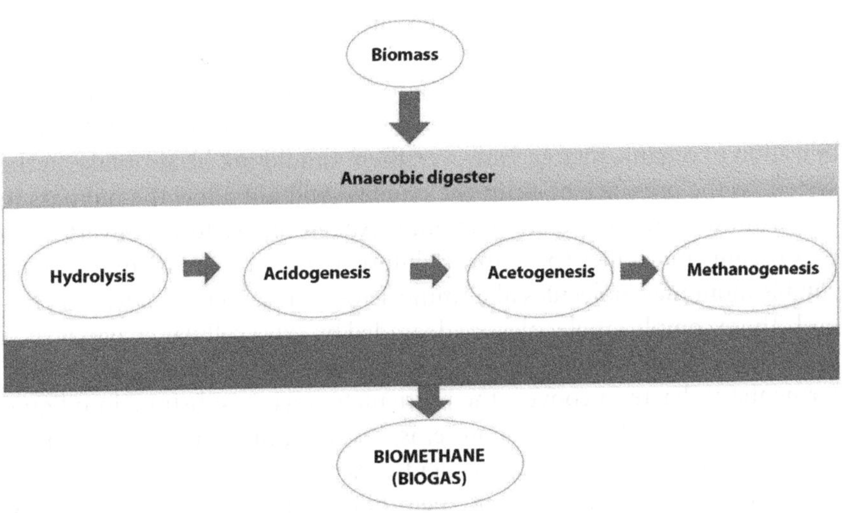

Figure 3.2 Outline steps of the anaerobic digestion process.

3.3.1 Hydrolysis

Hydrolysis is the first stage in AD. It is accomplished by the digestion and degradation of biopolymer particulate organic compounds and colloidal wastes into soluble monomeric or oligomeric organic compounds (Gerardi, 2003). This results in the putrefaction of complex organic polymeric materials such as carbohydrates, proteins, and lipids. Later, the compounds are hydrolyzed into smaller, water-soluble compounds such as sugars, amino acids, and long chain fatty acids using enzymes obtained from fermenting bacteria (microorganisms). The product of hydrolysis is a simple organic compound, which subsequently undergoes the process of absorption and degradation in the next step, acidogenesis by means of distinctive, facultative and obligate anaerobic bacteria. Because of this, short chain volatile fatty acids (VFA) are produced. Later, these VFAs are converted to acetate, H_2, and CO_2 by combining with alcohols. This phase involves hydrolyzing polysaccharides into monosaccharides, fats into glycerine, and fatty acids and proteins into amino acids. The enzymatic catalysis accelerates the hydrolysis process through oxidation of organic matter via a process called aerobic biological processes. The hydrolysis and aerobic degradation process is a rapid process, and the biogas produced is transformed into CO_2 from oxygen. When the substrate has been hydrolyzed, it becomes available for cell transportation and the fermentative bacteria can then degrade these substrates during the acidogenesis stage. Optimization of the hydrolysis process is, however, important to prevent inefficient degradation of the macromolecules, which could impact negatively on the rate of digestion or other biological activities, and consequently the biogas yield. It is therefore important to make sure that the culture of microorganisms is actively operational to allow the second process (acidogenesis) to take place. Physicochemical treatments can also be used to promote the solubilization of organic matter. However, there should not be air intake in the system as the presence of air in the biomass will not allow the biomass to perform their duties as anaerobic units. Anaerobic bacteria convert complex organic molecules (proteins, cellulose, lignin and lipids) into simpler soluble monomer molecules like amino acids, glucose, fatty acids, and glycerol. These complex molecules are degraded by extra cellular enzymes such as cellulase, proteases and lipases. Hence, in the first stage of hydrolysis, fermentative bacteria convert the particulate organic substrate into liquid monomers and polymers, i.e. proteins, carbohydrates, and fats to amino acids, monosaccharides, and fatty acids, respectively. Interventions of various microorganisms for biomethane production are shown in Figure 3.3.

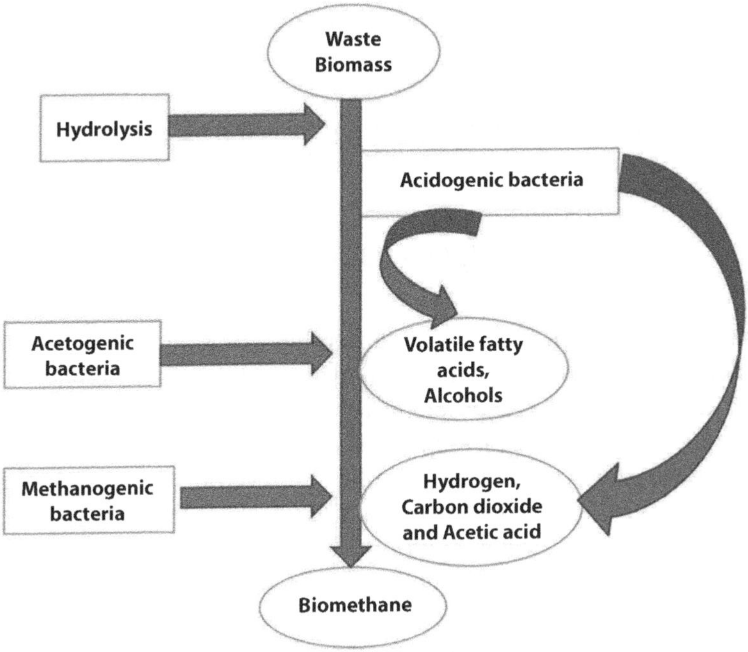

Figure 3.3 Microbial interventions during the biomethane production.

3.3.2 Acidogenesis

Acidogenesis is a complex process where anaerobic microorganisms degrade organic compounds to simpler low molecular organic acids. Some of the compounds produced by acidogens from glucose are acetate, lactate, succinate, ethanol, butanol, and acetone. Certain polysaccharides are converted to monosugars, proteins to amino acids, and fats to fatty acids and glycerol. Under high hydrogen partial pressure, acetic acid formation is lowered, and the substrate is converted to propionic acid, butyric acid, and ethanol. In the second stage of acidogenesis, acidogenic bacteria transform the products of the first reaction into short chain volatile acids, ketones, alcohols, H_2 and CO_2. The principal acidogenesis stage products are propionic acid (CH_3CH_2COOH), butyric acid ($CH_3CH_2CH_2COOH$), acetic acid (CH_3COOH), formic acid ($HCOOH$), lactic acid ($C_3H_6O_3$), ethanol (C_2H_5OH), and methanol (CH_3OH). The monomers produced from the hydrolysis process are further converted by a large diversity of facultative anaerobes and anaerobes through many fermentative pathways. Three acidogenesis reactions where glucose is converted to ethanol, propionate and acetic acid, respectively, are shown in Equations 3.1 and 3.2 (Ostrem, 2004), and Equation 3.3 (Bilitewski *et al.*, 1997).

$$C_6H_{12}O_6 \leftrightarrow 2CH_3CH_2OH + 2CO_2 \tag{3.1}$$

$$C_6H_{12}O_6 + 2H_2 \leftrightarrow 2CH_3CH_2COOH + 2H_2O \tag{3.2}$$

$$C_6H_{12}O_6 \rightarrow 3CH_3COOH \tag{3.3}$$

Various facultative and obligate anaerobes are explored for converting simpler molecules from the hydrolyzed macromolecules such as butyric acid, acetic acid, lactic acid, and propionic acids along with H_2, CO_2 and alcohols (Adekunle and Okolie, 2015; Kangle *et al.*, 2012). During this process, glucose breaks down into pyruvate via the Embden–Meyerhof–Parnas (EMP) pathway, which is later converted into Acetyl-CoA. After that, acetyl-CoA proceeds for either acetic acid or butyric acid formation. Phosphotransacetylase (PTA) helps in the conversion of acetyl-CoA to acetyl phosphate, which is later converted into acetic acid by acetate kinase (AK).

Whereas during butyrate forming pathway, phosphotransbutyrylase (PTB) facilitates the conversion of butyryl-CoA to butyryl phosphate, which is later converted into butyric acid by butyrate kinase (BK). In other types of the fermentation process of propionic acid formation, it is carried out by propionic acid bacteria (PAB) that converts glucose to phospoenol pyruvate and then to pyruvate via the EMP pathway. Further, pyruvate and methyl malonyl CoA react to form propionyl-CoA and oxaloacetate. During subsequent processes, propionyl-CoA converts to propionic acid.

3.3.3 Acetogenesis

During this process, products such as propionic acid, butyric acid, and alcohols are transformed by acetogenic bacteria into H_2, CO_2 and acetic acid. Hydrogen plays a transitional role in this process as the reaction will only occur if the hydrogen partial pressure is low enough to thermodynamically allow for conversion of all acids. The partial pressure is carried out by many H_2-scavenging bacteria. Hence, H_2 concentration of a digester is an indicator of their health (Mata–Alvarez, 2003). Equation 3.4 represents the conversion of propionate to acetate whereas Equations 3.5 and 3.6 show the conversion of glucose and ethanol to acetate during the third stage of anaerobic fermentation (Ostrem, 2004).

This process is governed by the partial H_2 pressure generated during acidogensis. If the partial pressure of H_2 is not too low, the conversion to acetic acid and CO_2 will not be thermodynamically favorable (Beguin and Aubert, 1994). This process occurs naturally in the presence of sulfate reducer. Hence, with the help of sulfate-reducing bacteria, the excess H_2 must be transferred to sulfate to produce H_2S. Similarly, H_2 is also transferred to CO_2 to form acetic acid with the help of acetogenic bacteria (Beguin and Aubert, 1994). A major amount of acetic acid is produced during acetogenesis. The CO_2 formed during this process is spent as the terminal electron acceptor, and H_2 acts as the electron donor so that later reduces CO_2 to acetic acid via the Wood–Ljungdahl pathway (Equation 3.4; Muller, 2003). Production of acetic acid is taken place in various ways as shown in Equations 3.5, 3.6, and 3.7.

$$2CO_2 + 4H_2 \rightarrow CH_3COOH + 2H_2O \tag{3.4}$$

$$CH_3CH_2COO - +3H_2O \leftrightarrow CH_3COO - +H + +HCO_3 - +3H_2 \tag{3.5}$$

$$C_6H_{12}O_6 + 2H_2O \leftrightarrow 2CH_3COOH + 2CO_2 + 4H_2 \tag{3.6}$$

$$CH_3CH_2OH + 2H_2O \leftrightarrow CH_3COO - +2H_2 + H + \tag{3.7}$$

3.3.4 Methanogenesis

This is the final step where CH_4 is formed by anaerobic respiration. In the presence of oxygen, methanogens growth is inhibited. There are two pathways through CO_2, and acetic acid acts as terminal electron acceptors for CH_4 production (Equations 3.8 and 3.9).

$$CO_2 + 4H_2O \rightarrow CH_4 + 2H_2O \tag{3.8}$$

$$CH_3COOH \rightarrow CH_4 + CO_2 \tag{3.9}$$

Methanogens like *Methanobacter* spp. and *Methanosaeta* spp. are explored for methanogenesis processes. Methanogens can only utilize H_2, formate, acetate, 2- propanol and 2-butanol for production of CH_4. In the fourth stage of methanogenesis, microorganisms convert the H_2 and acetic acid formed by the acid formers to CH_4 gas and CO_2 (Equations 3.10, 3.11, 3.12) (Verma, 2002). The bacteria responsible for this conversion are called methanogens, which are anaerobic in nature.

Methanogens are nothing but various forms of extremophiles. Thermophilic methanogens can grow at high (40°C–100°C) temperatures. Some mesophiles methanogens can perform their function in a temperature between 20°C and 40°C (Garcia *et al.*, 2000). On the other hand, psychrophiles grow at low temperatures (Nozhevnikova *et al.*, 2003); halophiles grow at increased salt concentration (Riffat and Krongthamchat, 2006); acidophiles can grow in acidic environments (Zhou and Nanqi, 2007); alkaliphiles can grow at high pH (Thakker and Ranade, 2002), and radophiles can grow in high radiation environments. Thermophilic methanogens are found in hot springs areas at a temperature range of 40°C–100°C, whereas extreme thermophilic methanogens are generally found in marine hydrothermal vents where temperatures go up to 350°C. Some thermophilic methanogens like *Methanothermobacter*, *Methanothermus*, *Methanothermococcus*, *Methanocaldococcus*, *Methanotorris*, and *Methanopyrus* grow at 80°C and above (Serranosilva *et al.*, 2014).

Methane-producing microbes are present in anaerobic environments like marine sediments, ruminant stomachs, and marshes. Being obligatory anaerobes, they are very sensitive to environmental changes. Unlike acidogenic and acetogenic bacteria, these belong to the archaea group having heterogeneous morphology, which distinguish them from other halophilic methanogens such as *Methanohalobium*, *Methanohalophilus* and *Methanolobus* that grow at 4.3mol/ L NaCl (Nazaries *et al.*, 2013). The general pH range for methanogenic bacteria is 6.0–8.0, and acidophiles can grow in the pH range of 5.6–6.2 whereas alkaliphiles perform their function at a pH range of 8.6–9.2 (Garcia *et al.*, 2000). Some methanogens are *Methanocaldococcus vulcanius*, *Methanothermobacter marburgensis*, *Methanococcoides burtonii*, *Methanosarcina acetivorans*, *Methanosphaerula palustris*, *Methanococcus maripaludis*, *Methanosaeta thermophile*, *Methanoregula boonei*, *Methanocaldococcus infernus*, *Methanothermus fervidus*, *Methanohalophilus mahii*, *Methanohalobium evestigatum*, *Methanocaldococcus fervens*, *Methanoplanus petrolearius*, *Methanospirillum hungatei*, *Methanocorpusculum labreanum*, *Methanosarcina mazei*, *Methanosarcina barkeri*, *Methanocella paludicola*, *Methanobrevibacter ruminantium*, *Methanocaldococcus jannaschi*, *Methanopyrus kandleri*, *Methanothermobacter thermautotrophicus*, and *Methanococcus vannielii*. The formation of various compounds during the entire biomethane generation are shown in Figure 3.4. Methane-producing bacteria convert low molecular weight organic compounds to CH_4 as the end product utilizing H_2, CO_2 and acetic acid.

Methanogenesis is the important step in AD (De Vrieze *et al.*, 2012) as approximately 70% of CH_4 used in the AD process is generated from methanogenesis. During this stage, CO_2-reducing and H_2-oxidizing methanogens convert H_2 and CO_2 to get methane, whereas acetoclastic methanogens make use of acetate to produce CH_4. Methanogens (archaea) utilize acetate, H_2 and CO_2, and in a very lesser amount methanol, methylamines, and formate, to produce CH_4 and CO_2. Now, these end products will become primary substrates for the methanogenic bacteria to generate biogas, which normally comprise 50%–75% CH_4, 50%–25% CO_2, trace amounts of nitrogen, H_2, and hydrogen sulfide. Methanogenesis shows the scope of biological activities in an anaerobic system and the condition of digestion.

$$CO_2 + 4H_2 \rightarrow CH_4 + 2H_2O \tag{3.10}$$

$$2C_2H_5OH + CO_2 \rightarrow CH_4 + 2CH_3COOH \tag{3.11}$$

$$CH_3COOH \rightarrow CH_4 + CO_2 \tag{3.12}$$

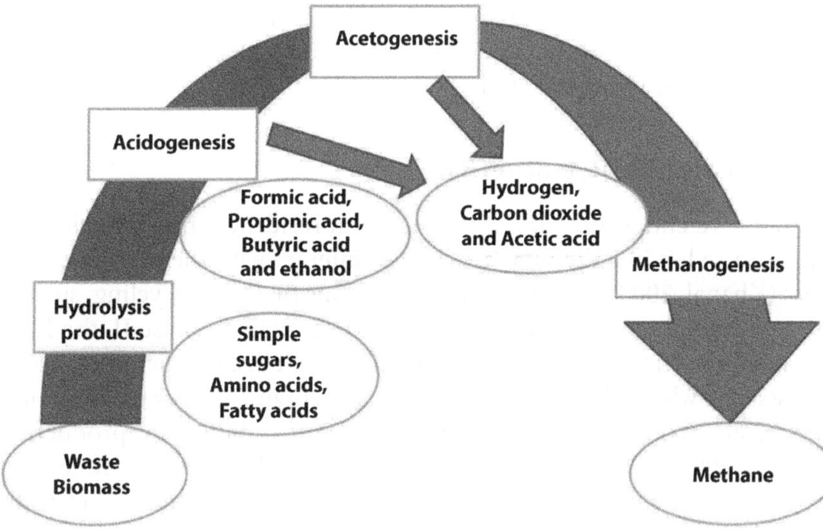

Figure 3.4 The production of various compounds during the biomethane generation.

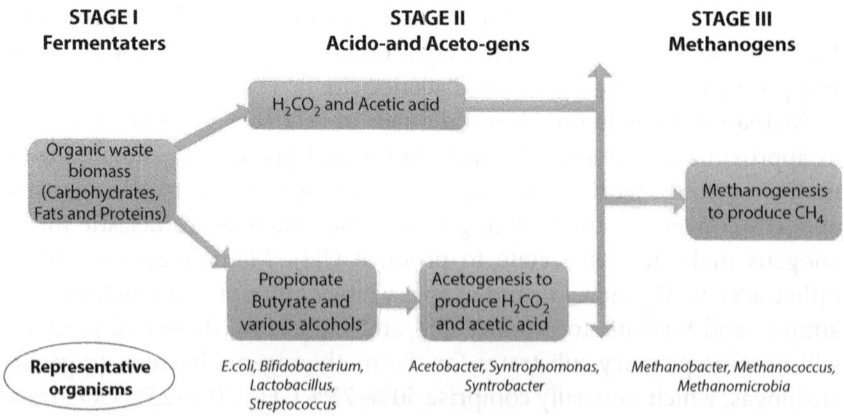

Figure 3.5 Processes and microorganisms during biomethane production.

Various microorganisms like acidogens and acetogens along with methanogens are employed for the production of biomethane performing different processes (Figure 3.5).

3.4 Main Factors Affecting the Biogas Production

The biogas production is controlled by many factors such as nutrients, pH of feedstock, temperature, flow rate of feed (loading rate), and retention time. If the values of these factors are not in a particular range, they may slow or stop biogas production (Angelidaki *et al.*, 2009). Some of the factors are presented in this section.

3.4.1 Hydraulic Retention Time

Hydraulic retention time (HRT) is the mean residence time for solid and liquid waste left in a digester (reactor) to interact with the microbial biomass (Khanal, 2008a). In flow-through systems that lack recycling, such as the continuous stirred tank reactor (CSTR) adopted in Phase II, the HRT and retention time of the microbial biomass or sludge are identical. In circumstances where the influent streams carry high solids concentrations, they require longer retention time to maximize bioenergy production (Khanal, 2008a). The HRT can be understood as the treatment time for a waste that undergoes AD, the higher the HRT the higher the removal efficiency because the biomass has enough time to be in close contact with the waste, therefore removing high amounts of contaminants from the waste being treated.

3.4.2 Nutrients

The inadequate availability of nutrient concentration in energy crops has resulted in issues such as low CH_4 production, acidification, and process instability in crop mono digestion, leading to implementation of low organic loading rates (OLRs) and long HRTs (Lebuhn *et al.*, 2008; Weiland, 2010). They affect the performance and solidity of the AD process (Hinken *et al.*, 2008; Lebuhn *et al.*, 2008; Scherer *et al.*, 2009). The above-mentioned setbacks indicate that sufficient amounts of both macro- and micronutrients (Bruni *et al.*, 2010) are significant for continuous performance of the biogas process.

3.4.3 pH of Feed Stock

The pH value of the matter is one of the essential factors. Methanogenic bacteria are sensitive to acidic conditions. This acidic condition could adversely affect the bacterial growth and CH_4 production. Different optimal pH values are reached at different stages of the AD process. These changes occur during a biological transformation that usually happens in different stages of the AD process. The lowest pH value can be as low as 5 during organic acid production, which takes place during the acetogenesis stage. The optimum range of pH for getting utmost biogas production in AD is 6.5–7.5 (Liu *et al.*, 2008), and this range of pH is comparatively broader in the plants. Several factors such as the substrate used and the digestion technique could vary the optimal value of the pH. For this reason, a constant pH level is of great importance, and to maintain a constant pH level, equilibrium buffers such as calcium carbonate or lime must be contributed into the system. Briefly, pH is a critical indicator in the anaerobic process. It provides a clear indication of the performance of the system, including the stability of the digestion. A lower pH indicates system failure or low buffering capacity and can inhibit digestion. High pH can also limit the methanogenesis process. The pH value is dependent on the following factors: VFA and bicarbonate concentration, the alkalinity of the system, and the amount of CO_2 in digester gas. The relationship between the VFA and bicarbonate concentration is very important to maintain a constant pH value within the system (Liu *et al.*, 2008).

3.4.4 Temperature

Anaerobic digestion normally functions within two different optimized temperature ranges, one at 35°C (mesophilic) and another at 55°C (thermophilic) (Davidsson *et al.*, 2008). Although thermophilic digestion has some advantages over mesophilic digestion, such as an improved reaction rate and pathogen reduction. In mesophilic digestion, microorganisms have comparatively lesser demand on nutrients (Takashima *et al.*, 2011), and mesophilic digestion can function like thermophilic digestion (Nges *et al.*, 2012). The temperature indicates the rate of biological reactions. It is a sensitive parameter that must be monitored regularly, especially amid weather changes. The choice of temperature (mesophilic or thermophilic) will depend on the type of expected outcome. However, the temperature should be suitable to the type of microorganisms used.

3.4.5 Organic Loading Rates

The amount of substrate (biomass) fed into the unit reactor system is called the OLR and is commonly expressed in terms of chemical oxygen demand kg/m3•day, volatile solids (VS) of total solids (TS)/ L•day or VS/m3•day. It has been reported that the AD of solid wastes in a single stage may encounter problems if the OLR is increased above the system capabilities and that the H_2 and the VFAs formed by the acidogenic bacteria are not consumed at the same rate by the methanogens. This is because acidogenic activity and the VFA intermediates produced in the acid forming stages triggers an increase in the acidogenic bacteria at higher OLRs, thereby reducing the growth of the methanogenic population. The increase in OLR and acidogenic activity (production of VFA, CO_2 and H_2) can result in an accumulation of organic acids and a decrease in pH and gas production. This in turn affects the biological activity of CH_4-producing methanogens as their growth is inhibited below a pH of 6.6, therefore reducing the production of CH_4, which is the main product of biogas. Therefore, determining the correct OLR for a particular substrate is critical for the optimization of reactor performance and maximizing CH_4 production. The CH_4 yield is generally measured by the amount of gas that can be produced per unit volume of VS contained in the feedstock after exposing it to AD for a sufficient time under a given temperature and other specific conditions (Zhang, 2012). The CH_4 yield is also an indication of the biodegradability of the substrate, as feedstock with low VS/TS, such as lignin, are not easily degraded using anaerobic processes. Therefore, the amount of gas produced is also very much substrate dependent.

3.4.6 Retention Time

A retention time should always be longer, which will render a greater degree of sludge stabilization and allow close contact between the biomass and the liquid flow during the process (Keay, 1981).

3.4.7 Mixing

In a mainstream anaerobic digester, mixing is normally remarked to increase CH_4 yields and to provide the digester more stability (Forday and Greenfield, 1983). Mixing brings a homogeneous environment and an efficient utilization of the entire digester volume. This can be attained by lowering hydraulic dead zones and preventing the development of unfavorable environmental conditions (low pH and high VFA). Mixing also assists in the elimination of extra CO_2, which has repressive effects at partial pressures larger than 0.2 atmospheres.

3.4.8 Oxygen

The presence of oxygen in any anaerobic reactor leads to a decrease in the digestion rate and can be very toxic to many anaerobic microbes. Though, it is also possible that before any emerging toxic effects, facultative anaerobes metabolize the dissolved oxygen (Zinder and Koch, 1984).

3.4.9 Volatile Fatty Acids

When there is a signal of high overloading of organic matter in the digester or during the starting up of the digester, high concentrations of VFAs are observed. They are normally related to toxicity and inhibitory results. It is understood that VFA suppression is because of their aggregation and subsequent reduction in pH, but it is also reported that some of the VFAs themselves are toxic to anaerobic microbes (Mara and Horan, 2003).

3.4.10 Free Ammonia

Free ammonia concentrations have the tendency for suppression at the amount above 100 mg/l, but the ionic form, NH+ 4, can only cause suppression at often higher concentrations, i.e. above 3000 mg/l.

3.4.11 Biogas Production

Anaerobic digestion follows four basic stages for producing biogas. Biogas production is the most productive solution for waste management, which has been accepted as the best alternative to renewable energy. The four stages followed for the process of biomethanation are hydrolysis, acidogenesis, acetogenesis, and methanogenesis. This process of AD is distinguished by the decomposition of organic matter into CH_4, CO_2, inorganic nutrients and compost in an anaerobic environment (Arsova, 2011; Ayu and Dyan Aryati, 2010).

Anaerobic digestion is a familiar process that converts organic waste to sustainable matter, biogas, and fertilizer under anaerobic conditions. The two primary methods to produce biogas from AD are named, wet AD and dry AD. These two methods carry a major difference that relates to the type of solid waste. Organic waste is handled by dry AD by simple mechanical sorting and with digestion taking place from waste in its solid form. On the other hand, wet AD makes waste turn into a homogenous pulp that can be pumped-up while being treated. The main composition of the biogas produced during the process of AD is CH_4 and CO_2, where 60%–65% is CH_4, which is flammable and is regarded as a substitute to traditional energy (Khanal, 2008b). As biogas utilization technology improves, it becomes more widely used, which is waste-to-energy (Khanal, 2008a). Biogas is also being used as fuel for cooking, lightning, and vehicles (Khanal, 2008a).

Anaerobic digestion is accomplished in an airtight cylindrical tank, called digester (Figure 3.6). A digester is a structure of concrete bricks and cement or steel. The sides have a wide opening (charge pit) for the entry of organic materials required for digestion. Just above the digester lies a cylindrical container for the collection of gas.

Biogas is a combustible gas, which has no color and is produced using the anaerobic process that is in the absence of oxygen by the biological breakdown of organic matter (Umeghalu *et al.*, 2012). The biogas occurs from "biogenic materials" (Umeghalu *et al.*, 2012), and it is produced from the AD of biodegradable substances such as biomass, cow dung, green waste, and agricultural residue such as cassava and sugar cane. Biogas is a mixture of different gases, mainly CH_4, CO_2, 1%–5% other gases, including H_2, because of which it is an attractive source of energy. The energy that liberates from biogas makes it a suitable fuel in any country for heating and cooking purposes. Biogas has the potential to be used in an anaerobic digester where the energy in the gas is converted into electricity and heat using a gas engine (Sorathia *et al.*, 2012).

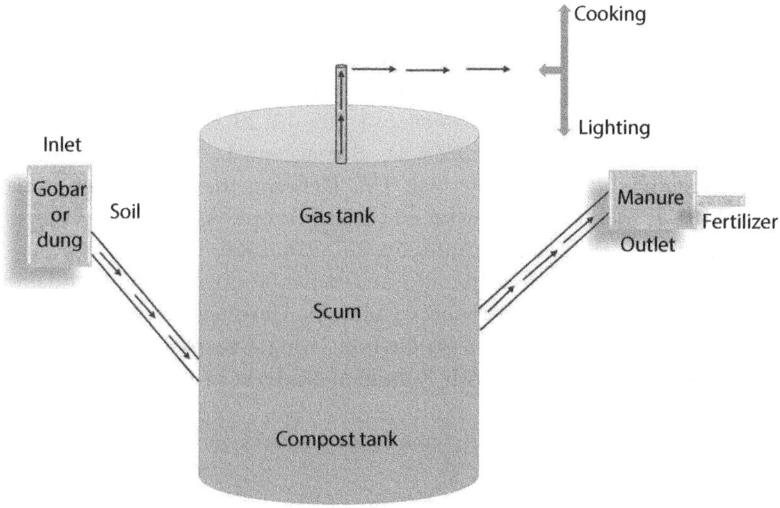

Figure 3.6 Diagram of a Vayu gas (gobar gas) plant.

3.5 Conclusion

The potential of CH_4 as an alternative global energy source is noteworthy as far as the growing energy demand and diminishing oil reserves are concerned. Various process technologies and advances in research on CH_4 have been explored. Worldwide increasing environmental CO_2 concentration which contribute to greenhouse gas emissions leads to global warming which can be converted to CH_4 to lessen the burden on the fossil-based fuel sources. In the context of global energy utilization from biobased sources, low cost technologies may be adapted for enhancing the production level of biofuels.

Bioconversion technologies act as a vital role for the conversion of renewable waste biomass into biofuels so that these can be focused on the framework of cost-benefit analysis having a higher yield. Biotechnological interventions and supplementation of genetic engineering aspects can evolve novel microorganisms, which can solve the problems of cost limitation problems. Exploring advanced methods of process integration by minimizing capital investment, maximizing energy efficiency, and subsequently improving overall economics, the major cost disadvantage of current second-generation biofuels can be overcome, thereby becoming intensive into the biofuel markets in the coming future. Hence, lignocellulosic waste biomass by green biotechnology will be the focus of future research.

References

Adekunle, K.F. and Okolie, J.A., A review of biochemical process of anaerobic digestion. *Adv. Biosci. Biotechnol.*, 6, 3, 205–12, 2015.

Angelidaki, I., Alves, M., Bolzonella, D., Borzacconi, L., Campos, J., Guwy, A., Kalyuzhnyi, S., Jenicek, P., Lier, J.V., Defining the biomethane potential (BMP) of solid organic wastes and energy crops: A proposed protocol for batch assays. *Water Sci. Technol.*, 59, 927–934, 2009.

Arsova, L., Department of earth and Environmental Engineering foundation of Engineering and Applied Science Columbia University, New York, 2010.

Ayu, O. and Dyan, A.V., Biogas Production Using Anaerobic Biodigester from Cassava Starch Effluent with Ruminant Bacteria as Biocatalyst, vol. 1–10, 2010.

Bailey, J. and Ollis, D., Applied enzyme catalysis. *Biochem. Eng. Fundamentals.*, 2, 157–227, 1986.

Beguin, P. and Aubert, J.P., The biological degradation of cellulose. *FEMS Microbiol. Rev.*, 13, 1, 25–58, 1994.

Bhatia, L., Sarangi, P.K., *et al.*, Vo, Sarangi, Current Advancements in Microbial Fuel Cell Technologies, in: *Biorefinery of Alternative resources: Targeting green fuels and platform chemicals*, S. Nanda, D.V.N., P.K. (Eds.), pp. 477–494, Springer Nature, Singapore, 2020.

Bruni, E., Jensen, A.P., Pedersen, E.S., Angelidaki, I., Anaerobic digestion of maize focusing on variety, harvest time and pretreatment. *Appl. Energ.*, 87, 2212–2217, 2010.

Chandra, R., Takeuchi, H., Hasegawa, T., Methane production from lignocellulosic agricultural crop wastes: A review in context to second generation of biofuel production. *Renew. Sust. Energ. Rev.*, 16, 3, 1462–1476, 2012.

Davidsson, A., Lövstedt, C., La Cour Jansen, J., Gruvberger, C., Aspegren, H., Co-digestion of grease trap sludge and sewage sludge. *Waste Manage.*, 28, 986–992, 2008.

De Vrieze, J., Hennebel, T., Boon, N., Verstraete, W., *Methanosarcina*: The rediscovered methanogen for heavy duty biomethanation. *Bioresour. Technol.*, 112, 1–9, 2012.

Eastman, J. and Ferguson, J., Solubilization of particulate organic carbon during the acid phase of anaerobic digestion. *J. Water Pollut. Control Fed.*, 8, 352–366, 1981.

Fang, C., Angelidaki, I., Boe, K., Biogas production from food-processing industrial wastes by anaerobic digestion. *Water Res.*, 90343–4, 1–59, 2010.

Forday, W. and Greenfield, P.F., Anaerobic digestion. *Effluent Water Treat. J.*, 23, 405–413, 1983.

Garcia, J.L., Patel, B.K., Ollivier, B., Taxonomic, phylogenetic, and ecological diversity of methanogenic archaea. *Anaerobe*, 6, 4, 205–226, 2000.

Gerardi, M., *The Microbiology of Anaerobic Digesters*, Wiley-Interscience, John Wiley and Sons, Inc.: New Jersey, Hoboken, 2003.

Hinken, L., Urban, I., Haun, E., Weichgrebe, D., Rosenwinkel, K., The valuation of malnutrition in the mono-digestion of maize silage by anaerobic batch tests. *Water Sci. Technol.*, 58, 1453–1459, 2008.

Holm-Nielsen, J.B., Al Seadi, T., Oleskowicz-Popiel, P., The future of anaerobic digestion and biogas utilization. *Bioresour. Technol.*, 100, 22, 5478–84, 2009.

Kalyuzhnyi, S., Veeken, A., Hamelers, B., Two-particle model of anaerobic solid state fermentation. *Water Sci. Technol.*, 41, 43–40, 2000.

Kangle, K.M., Kore, S.V., Kore, V.S., Kulkarni, G.S., Recent trends in anaerobic codigestion: a review. *Univers. J. Environ. Res. Technol.*, 2, 4, 210–9, 2012.

Keay, G., *Anaerobic digestion: Design and upgrading. Sewage Treatment Plants, Design, Operation and Upgrading*, pp. 1–18, Lancaster, PA: University of Pretoria, 1981.

Khanal, S.K., *Anaerobic Biotechnology for Bioenergy Production. Principles and Applications*, A John Wiley and Sons, USA, 2008a.

Khanal, S.K., Overview of anaerobic biotechnology, in: *Anaerobic Biotechnology for Bioenergy Production: Principles and Applications*, vol. 20, pp. 385–406, Wiley, New York, 2008b.

Kopyscinski, J., Schildhauer, T.J., Biollaz, S.M., *Production of synthetic natural gas (SNG)*, 2010.

Lebuhn, M., Liu, F., Heuwinkel, H., Gronauer, A., Biogas production from mono-digestion of maize silage-long-term process stability and requirements. *Water Sci. Technol.*, 58, 1645–1650, 2008.

Liu, C.F., Yuan, X.Z., Zeng, G.M., Li, W.W., Li, J., Prediction of methane yield at optimum pH for anaerobic digestion of organic fraction of municipal solid waste. *Bioresour. Technol.*, 99, 882–888, 2008.

Mara, D. and Horan, N.J., Title??, in: *Handbook of Water and Wastewater Microbiology*, Elsevier, Amsterdam, 2003.

Müller, V., Energy conservation in acetogenic bacteria. *Appl. Environ. Microbiol.*, 69, 11, 6345–53, 2003.

Nanda, S., Rana, R., *et al.*, A Spotlight on Butanol and Propanol as Next-Generation Synthetic Fuels, in: *Biorefinery of Alternative resources: Targeting green fuels and platform chemicals*, S. Nanda, D.V.N. Vo, P.K. Sarangi (Eds.), pp. 105–126, Springer Nature, Singapore, 2020.

Nazaries, L., Murrell, J.C., Millard, P., Baggs, L., Singh, B.K., Methane, microbes and models: fundamental understanding of the soil methane cycle for future predictions. *Environ. Microbiol.*, 15, 9, 2395–417, 2013.

Nges, I.A., Lovisa, B., Hinrich, U., *Anaerobic Digestion of Crop and Waste Biomass: Impact of Feedstock Characteristics on Process Performance*, Lund University, Denmark, 2012.

Nozhevnikova, A.N., Zepp, K., Vazquez, F., Zehnder Alexander, J.B., Holliger, C., Evidence for the existence of psychrophilic methanogenic communities in anoxic sediments of deep Lakes. *Appl. Environ. Microbiol.*, 69, 3, 1832–5, 2003.

Pasin, T.M., de Almeida, P.Z., *et al.*, Bioconversion of Agro-industrial Residues to Second-Generation Bioethanol, in: *Biorefinery of Alternative Resources: Targeting Green Fuels and Platform Chemicals*, S. Nanda, D.V.N. Vo, P.K. Sarangi (Eds.), Springer, Singapore, 2020.

Riffat, R. and Krongthamchat, K., Specific methanogenic activity of halophilic and mixed cultures in saline wastewater. *Int. J. Environ. Sci. Te.*, 2, 4, 291–9, 2006.

Sarangi, P.K., Singh, J., Singh, T.A., Pectin from pineapple waste: Isolation and Process Optimization. *Int. J. Curr. Microbiol. Appl. Sci.*, 9, 5, 143–148, 2020b.

Sarangi, P.K. and Nanda, S., Bioconversion of agro-wastes into phenolic flavour compounds, in: *Biotechnology for Sustainable Energy and Products*, P.K. Sarangi and S. Nanda (Eds.), pp. 266–284, IK International Publishing House, India, 2019a.

Sarangi, P.K., Nanda, S., Vo, Recent advances in consolidated bioprocessing for microbe-assisted biofuel production, in: *Fuel processing and energy utilization*, S. Nanda, P.K. Sarangi, D.V.N. (Eds.), pp. 141–157, CRC Press, Boca Raton, FL, 2019c.

Sarangi, P.K., Nanda, S., Sahoo, H.P., Maximization of vanillin production by standardizing different cultural conditions for ferulic acid degradation. *N. Y. Sci. J.*, 3, 7, 77–79, 2010.

Sarangi, P.K. and Nanda, S., Valorization of pineapple wastes for biomethane generation, in: *Biogas Technology*, S. Mishra, T.K. Adhya, S.K. Ojha (Eds.), pp. 169–180, New India Publishing Agency, New Delhi, India, 2019b.

Sarangi, P.K. and Nanda, S., *et al.*, Technological Advancements in the Production and Application of Biomethanol, in: *Biorefinery of Alternative resources: Targeting green fuels and platform chemicals*, S. Nanda, D.V.N. Vo, P.K. Sarangi (Eds.), pp. 127–140, Springer Nature, Singapore, 2020.

Sarangi, P.K., Sahoo, H.P., Ghosh, S., Mitra, A., Biotransformation of ferulic acid into vanillin by a *Streptomyces* isolate S10. *Asian J. Microbiol. Biotechnol. Environ. Sci.*, 11, 273–278, 2009b.

Sarangi, P.K. and Sahoo, H.P., Biotransformation of ferulic acid into value-added products. *Asian J. Microbiol. Biotechnol. Environ. Sci.*, 12, 3, 523–525, 2010a.

Sarangi, P.K. and Sahoo, H.P., Enhancing the rate of ferulic acid bioconversion utilizing glucose as carbon source. *J. Am. Sci.*, 6, 5, 115–117, 2010c.

Sarangi, P.K. and Sahoo, H.P., Ferulic Acid production from wheat bran using *Staphylococcus aureus*. *N. Y. Sci. J.*, 3, 4, 79–81, 2010b.

Sarangi, P.K., Sahoo, H.P., Ghosh, S., Mitra, A., Isolation of ferulic acid from wheat bran using *Streptomyces* isolates. *Int. J. Appl. Agric. Sci.*, 4, 1, 57–62, 2009a.

Sarangi, P.K., Singh, T.A., *et al.*, Pineapple as potential crop resource: Perspective and value addition, in: *Food Bioresources and Ethnic Foods of Manipur, Northeast, India*, pp. 83–91, Empyreal Publishing House, 2019.

Schink, B., Energetics of syntrophic cooperation in methanogenic degradation. *Microbiol. Mol. Biol. Rev.*, 61, 262–280, 1997.

Serrano-Silva, N., Sarria-Guzmán, Y., Dendooven, L., Luna-Guido, M., Methanogenesis and methanotrophy in soil: A review. *Pedosphere*, 24, 3, 291–307, 2014.

Siang, T.J., Roslan, N.A., *et al.*, Recent Advances in Steam Reforming of Glycerol for Syngas Production, in: *Biorefinery of Alternative resources: Targeting green fuels and platform chemicals*, S. Nanda, D.V.N. Vo, P.K. Sarangi (Eds.), pp. 399–426, Springer Nature, Singapore, 2020.

Sorathia, H.S., Rathod, P.P., Sorathiya, A.S., Bio-gas generation and factors affecting the bio-gas generation–a review study. *Int. J. Adv. Eng. Technol.*, 3, 72–78, 2012.

Speece, R.E., Sulfide production, in: *Anaerobic biotechnology for industrial waste-waters*, R.E. Speece, (Ed.), pp. 287–318, Archaea Press, Nashville, Tennessee, 1996.

Sutaryo, S., Optimisation and inhibition of anaerobic digestion of livestock manure. *Tech. Rep. Biol. Chem. Eng.*, 1.3, 2012.

Takashima, M., Shimada, K., Speece, R.E., Minimum requirements for trace metals (iron, nickel, cobalt, and zinc) in thermophilic and mesophilic methane fermentation from glucose. *Water Environ. Res.*, 83, 339–346, 2011.

Thakker, C.D. and Ranade, D.R., An alkalophilic *Methanosarcina*isolated from Lonar crater. *Curr. Sci.*, 82, 4, 455–8, 2002.

Umeghalu, C., Chukwuma, E., Okonkwo, I., Umeh, S., Potentials for biogas production in Anambra State of Nigeria using cow dung and poultry droppings. *Int. J. Vet. Sci.*, 1, 26–30, 2012.

Weiland, P., Biogas production: Current state and perspectives. *Appl. Microbiol. Biotechnol.*, 85, 849–860, 2010.

White, J., *Biogas generation potential of coconut copra in the anaerobic digestion process*, 2011.

Zhang, P., *Biogas-Its Potential as an Energy Source In Rural Households With Particular Emphasis On China*, Norwegian University of Life Science, Akershus, 2012, MSc Thesis.

Zhou, X. and Nanqi, R., Acid resistance of methanogenic bacteria in a two-stage anaerobic process treating high concentration methanol wastewater. *Front. Environ. Sci. Eng. China*, 1, 1, 53–6, 2007.

Zinder, S.H. and Koch, M., Non-aceticlastic methanogenesis from acetate: Acetate oxidation by a thermophilic syntrophic coculture. *Arch. Microbiol.*, 138, 263–272, 1984.

4

Production of Biohydrogen: Process Technologies at a Glance

Prakash Kumar Sarangi

College of Agriculture, Central Agricultural University, Imphal, Manipur, India

Abstract
The demands for energy gained a promising and rapid enhancement due to industrial revolution and global population growth. Major contributions of energy use are shared by fossil-based fuel sources, which not only produce environmental pollutants but are also in limited supply. Hence, there is a need for a green, sustainable, and alternative fuel generation that can mitigate the problems associated with the massive use of fossil fuels. Biobased resources can be used as alternative and sustainable material for future fuel production. Among various biofuels, biohydrogen is seen as potential fuel for mitigating the future energy crisis, thereby environmental concerns linked with fossil fuels. Various processes, like thermochemical and biological, are implemented to produce hydrogen. Dark fermentation is shown to have the most potential, and this chapter summarizes and discusses it. Process technologies and future possibilities associated with biohydrogen production are described in this chapter.

Keywords: Biohydrogen, fossil fuel, microorganisms, sustainable, energy

4.1 Introduction

The global need for energy is escalating due to population growth and industrial development. Presently, fossil fuels, such as oil, coal, and gas, contribute greatly to the energy market (Ghimire *et al.*, 2015). Dependence on energy in various sectors is caused by fossil-based fuels, which will be exhausted rapidly in decades. These energy sources generate pollutants, which are the main cause for air pollution. The adverse effects of fossil fuel combustion has forced academicians and researchers globally to search for

Prakash Kumar Sarangi. Waste Biomass to Alternative Fuels, (67–78) © 2025 Scrivener Publishing LLC

alternative sources of energy capable of substituting conventional sources. Presently, the establishment of environmental sustainability with clean and green energy carriers is seen as a major concern. In this context, biobased renewable feedstocks have the potential towards for energy, fuel, and chemical production, thereby maintaining sustainability.

Waste biomass, having food and agro-based origins, can result in a worldwide threat to public health as far as environmental issues are concerned, having been used in animal feed (Min *et al.*, 2011). Global academic and industrial researchers are currently focusing on the extraction of high-value fuels and biochemicals from these waste biomass (Sarangi *et al.*, 2020a; Sarangi *et al.*, 2020b). Agro-industries generate various by-products including corn cobs, pineapple waste, sugarcane bagasse, bran, and straw from wheat and rice, sugar beet pulp, and coffee pulp, which have the potential as cheap substrates for biological degradation to obtain fuels and value-added products using environment-friendly bioprocess technology and biotechnological interventions (Sarangi *et al.*, 2017; Sarangi *et al.*, 2019; Sarangi and Nanda, 2019a; Sarangi and Nanda, 2019b; Sarangi and Nanda, 2019c). A huge accumulation of waste materials has great perspectives for value addition into various biochemicals for different industries, being carbohydrates and phenolic in nature (Subudhi *et al.*, 2023; Sarangi *et al.*, 2023; Sarangi *et al.*, 2023a; Sarangi *et al.*, 2023b). Waste bioconversion is an increasingly widespread alternative for the treatment of agricultural and industrial waste for economic and ecological advantages. Lignocellulosic biomass (LB) acts as a renewable energy source, attracting continuing efforts to produce biochemicals for future uses. Hence, attention is geared towards new green, clean, and sustainable renewable technologies to avoid complete dependence on fossil-based products (Sarangi and Nanda, 2018; Sarangi *et al.*, 2017; Sarangi, 2019). The limited nature of fossil-based fuels forces global researchers to find green energy production technologies with environmental sustainability (Sarangi *et al.*, 2020).

The shifts of the biorefinery processes toward the production of fuels and chemicals through renewable sources are targeted, which is the biggest economic challenge among researchers and academicians globally. A wide array of feedstocks can be applied to produce energy and fuel sources for the mitigation of an energy crisis (Nanda *et al.*, 2020; Pasin *et al.*, 2020; Siang *et al.*, 2020; Bhatia *et al.*, 2020; Sarangi *et al.*, 2020). Therefore, novel bioprocess technologies along with new biotechnological tools may be focused towards new green and renewable technologies for safer future environments, overcoming much dependence on petroleum-based products (Sarangi and Nanda, 2018; Sarangi *et al.*, 2017).

The share of energy from renewable sources in 2011 encompasses about 20.5% of the total energy sector, which was 16% in 2009. Since the last few decades, there has been a significant increase in the production of renewable energy sources like solar, tidal, geothermal, wind, and biomass (Renewables, 2011). The harnessing of waste biomass in this regard gained a massive global focus towards its conversion into different biofuels and bioenergy (Sarangi and Nanda, 2019; Sarangi *et al.*, 2020; Nanda *et al.*, 2020; Siang *et al.*, 2020; Bhatia *et al.*, 2020).

Hydrogen among all biofuels has put many applications in fuel and energy sectors. It is made suitable for future sustainable fuel sources. The production of hydrogen has been found from both fossil and waste biomass. The estimation showed that hydrogen production shared about 11% of the entire renewable energy by 2025 and an increase to about 34% out of total renewable energy share by 2050. (Balat and Kırtay, 2010). This chapter describes the production technologies of biohydrogen by dark fermentation and its process technologies. Advanced technologies and future possibilities during dark fermentation were also described.

4.2 Applications of Hydrogen

The increasing interest in hydrogen production and further utilization as a fuel source has been detected since the last decade. Having great potential as an energy and fuel source, hydrogen can be termed as the fuel of the future, being used as a fuel source in internal combustion or jet engines along with fuel cells technology. It has high net calorific value as compared to other important fossil-based fuels. It has been detected that one kilogram of hydrogen shows an equivalent to 2.75 kg of gasoline and 6 kg of methanol (Grimes *et al.*, 2008). Combustion of hydrogen produces water and heat energy, thereby positioning as the environmentally friendly fuel having versatile applications towards energy and fuel sector (Table 4.1) Although it has several advantages in fuel and energy sectors over other synthetic biofuels, it also has limitations in its explosivity nature when reacted with oxygen. Its low storage boiling point is also another limitation issue, which is difficult compared to have other options. Hydrogen is considered as a promising renewable green energy source for a sustainable future, having many peculiar characteristics (Figure 4.1). The production of hydrogen from biomass has gained momentum globally due to many environmental and socio-economic advantages (Kotay and Das, 2008).

Biohydrogen has great potential in reducing carbon emissions. The application of hydrogen has been extended in present and future times.

Table 4.1 Important attractive features of hydrogen as an alternative fuel source.

Sl. no.	Characteristics	References
1	About 2.75 times greater energy content than conventional fossil-based fuels, amounting to 122KJ/g.	Faloye *et al.*, 2014
2	The combustion of hydrogen results into water vapor being non-greenhouse gas and ecofriendly.	Armor, 2005
3	As compared to gasoline, the combustion of hydrogen is considered nearly 50% more efficient in automobiles. A hydrogen battery is more efficient than other fuel and energy sources for automobiles.	Reith *et al.*, 2003
4	The use of hydrogen may directly be in combustion engines or toward electricity generation with fuel cell technologies.	Alves *et al.*, 2013

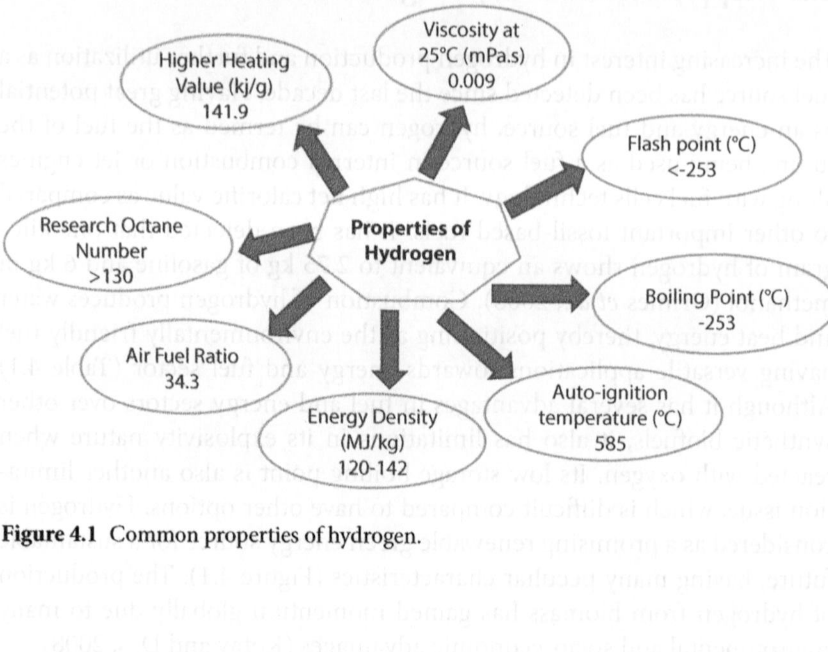

Figure 4.1 Common properties of hydrogen.

It is regarded as one of the most competitive, alternative energy and fuel sources. The largest user of hydrogen fuel is associated with fertilizers and petroleum-based industries, having 50% and 37% share, respectively. Apart from being a fuel and energy source, hydrogen is used as the main application in starting materials in various industries such as chemical, refinery, and pharmaceutical sectors. It also has major applications in the synthesis of ammonia and methanol, along with hydrogenation of liquid oils (Grimes *et al.*, 2008). Furthermore, the applications of hydrogen gains momentum in the synthesis of aniline from nitrobenzene and the synthesis of hexamethylenediamine (Grimes *et al.*, 2008). In food industries, hydrogen is used in the hydrogenation of fats and oil. Due to its safety and flexibility in applications toward all economy dimensions, hydrogen has the potential toward becoming "the world's clean energy source," being the first carrier of energy, thereby establishing a strong foundation to the significant energy system (Show *et al.*, 2012).

4.3 The Production of Hydrogen

Due to the non-occurrence of hydrogen as an elemental form in nature rather than chemical compounds, it is required to convert those compounds to a generation of hydrogen. Although the presence of hydrogen is in abundance, the production of its pure form needs various conversion technologies like thermochemical, steam reforming of natural gas and petroleum, photocatalysis, fermentation in presence (photo-fermentation), and the absence of light (dark fermentation), along with microbial electrolytic cells.

Presently, the production of hydrogen is completely dependent on conventional fossil-based sources by the process of steam reforming, the gasification of coal, catalytic decomposition of natural gas, and the partial oxidation of heavy hydrocarbons fraction of petroleum (Figure 4.2). These are cost-bearing processes, hence the search for non-polluting sources. Major limitations, such as being energy-intensive and having high temperatures, along with the production of toxic pollutants force for searching cost-effective alternative sources toward the sustainable production of hydrogen. The major environmental polluting oxides along with radioactive substances are generated due to such methods, creating bottlenecks for hydrogen production through conventional ways.

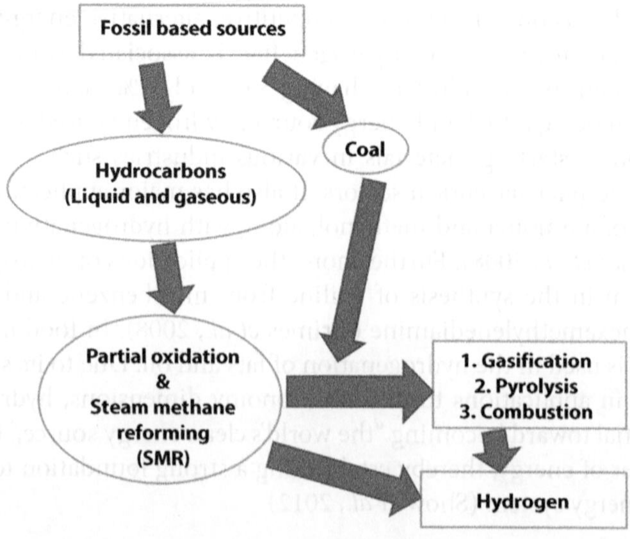

Figure 4.2 The production of hydrogen from fossil-based sources.

Technologies like electrochemical methods such as electrolysis and photolysis, as well as biochemical methods such as dark fermentation and photo-fermentation can also be explored as alternative sources for hydrogen production. Hydrogen production can be classified into two main categories: non-renewable fossil fuel energy sources (coal, natural gas, and other hydrocarbons) and renewable energy sources such as biomass and solar. Though the production of hydrogen through the electrolysis of water requires electrical energy, which comes from coal or natural gas sources, the major benefits about the non-emission of carbon dioxide are detected.

Transformation technologies toward hydrogen generation depend on the type of energy sources. Thermochemical conversion is generally employed when fossil fuels are used, but alternative technologies like gasification, pyrolysis, fermentation, or electrolysis can be explored for the conversion of waste biomass to produce hydrogen. Thus, to transform organic waste biomass, both thermochemical, as well as biochemical technologies, can be implemented based on the biomass type. Waste biomass, having high lignocellulosic content more energy-yielding value, can be directly converted to hydrogen through thermochemical and biochemical methods. Biochemical methods are particularly suitable for biomass, having more biodegradability level and low crystallinity content to produce hydrogen. Electrochemical methods are extensively used to generate hydrogen through the electrolytic conversion of water. With the help of biological involvements, renewable organic waste biomass, byproducts from agro-industrial process and water may be focused to produce hydrogen as an environmentally friendly and less energy-intensive energy carrier (Figure 4.3).

Due to the presence of LB in an abundance manner, various strategies are employed for the utilization of these waste biomass toward biohydrogen production [Rezania *et al.*, 2017]. Various lignocellulosic materials like rice straw, rice husk, sugarcane bagasse, corn stalk, and corn stover can be used for hydrogen production through dark fermentation, as well as microbial electrolysis cells methods (Kadier *et al.*, 2014; Singh *et al.*, 2015). Moreover, pretreatment methods are necessitated for better availability of sugar compounds toward enzymatic degradation for biohydrogen production.

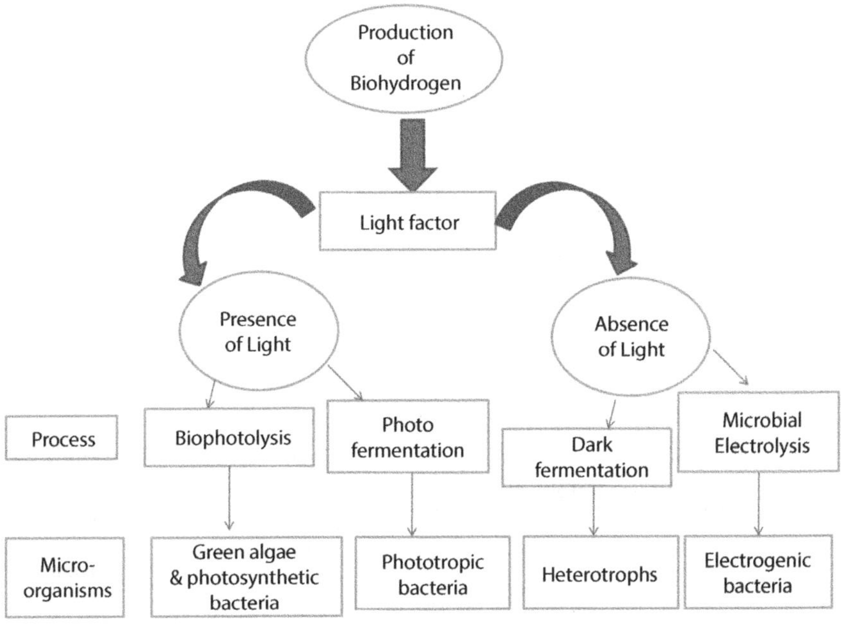

Figure 4.3 The biological production of biohydrogen.

4.4 Biochemistry of Dark Fermentation

The production of biohydrogen is accomplished by the mechanism of fermentations via glycolysis from glucose through various metabolic processes (Figure 4.4). Glycolysis is the first step of fermentation in which glucose is converted to pyruvate along with the formation of the reduced form of one molecule of nicotinamide adenine dinucleotide and two molecules of adenosine triphosphate. Through the decarboxylation process, pyruvate can be converted to acetyl-CoA catalyzed by pyruvate ferredoxin oxidoreductase. Through this reaction, in addition to acetyl-CoA and CO_2, reduced ferredoxin helps in the reduction of [FeFe]-hydrogenases that subsequently reduce protons yielding hydrogen. On the other hand, another possible pathway with the formation of formate catalyzed by pyruvate formate lyase in which formate is converted to hydrogen and carbon dioxide in the presence of [NiFe]-hydrogenases or [FeFe]-hydrogenases. Finally, acetyl coenzyme A can also be converted into a wide range of value-added products such as ethanol, butanol, acetone, butyric acid or acetic acid (Hawkes *et al.*, 2007; Hallenbeck, 2012). Enhanced production strategies are shown in Figure 4.5.

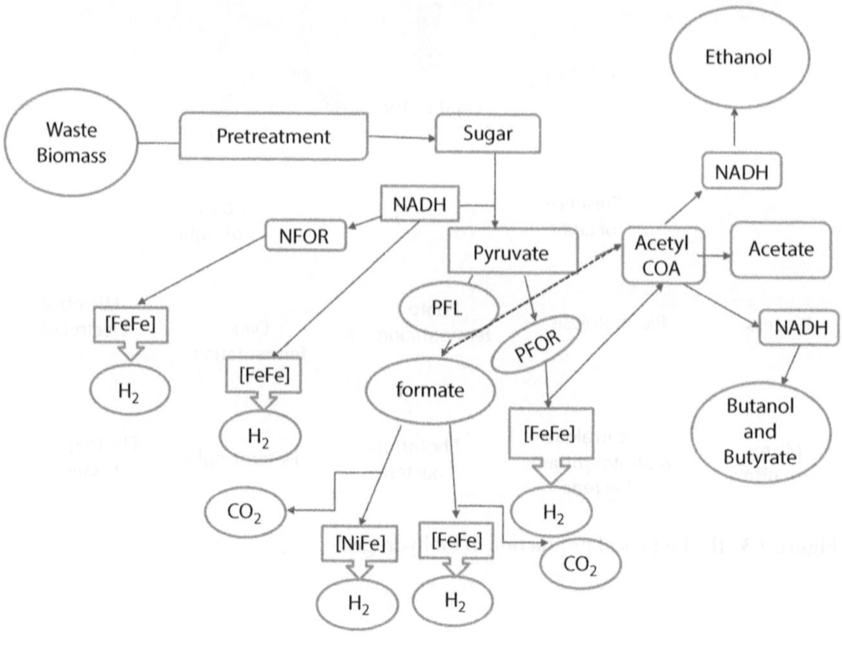

Figure 4.4 Mechanism of dark fermentation.

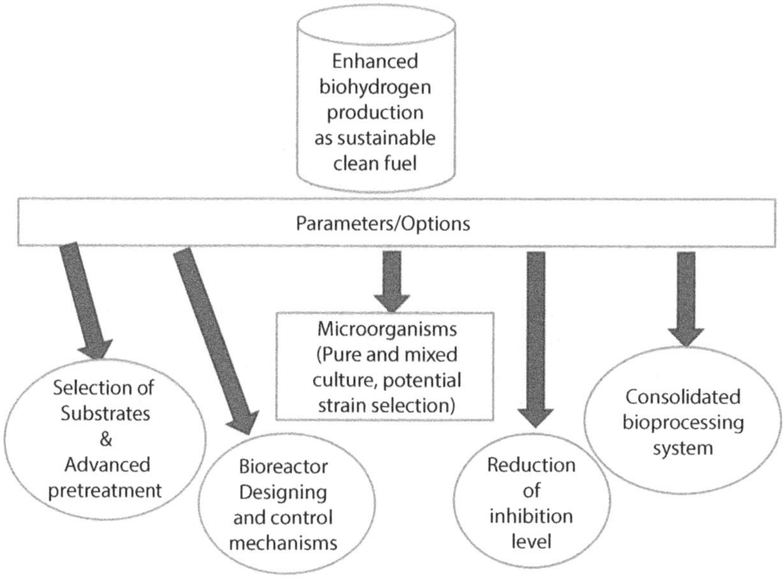

Figure 4.5 Strategies toward an enhanced production of biohydrogen.

4.5 Conclusion

The biggest concern and challenge of the bioenergy sector is searching for a green and renewable source that must be adopted ecologically and economically. Biohydrogen generation by implementation of dark fermentation is the promising substitute for the development of sustainable renewable processes. Harnessing waste biomass toward hydrogen production through this method is regarded as the most capable way by utilizing microbial biomass. The utilization of microbial diversities and their exploitation into hydrogen production along with optimization of various process parameters are the few challenges for maximum recovery of biohydrogen to be utilized as future biofuel. Various parameters such as temperature, pH, hydraulic retention time, and partial pressure of hydrogen affect the recovery of hydrogen. Other factors such as nature and the type of microbial biomass, pretreatment process, growth medium, cultural conditions, characteristics, and the concentration of feedstock also have vital roles in controlling the final recovery of biohydrogen. Bioreactor designing and the mode of operation also cannot be ignored as far as the production of hydrogen is concerned.

References

Alves, H.J., Bley, Jr., C., Niklevicz, R.R., Frigo, E.P., Frigo, M.S., Coimbra-Araújo, C.H., Overview of hydrogen production technologies from biogas and the applications in fuel cells. *Int. J. Hydrog. Energ.*, 38, 5215–25, 2013.

Armor, J.N., Catalysis and the hydrogen economy. *Catal. Lett.*, 101, 3–4, 131–135, 2005.

Balat, H. and Kırtay, E., Hydrogen from biomass—present scenario and future prospects. *Int. J. Hydrog. Energ*, 35, 7416–7426, 2010.

Bhatia, L., Sarangi, P.K., Nanda, S., Current Advancements in Microbial Fuel Cell Technologies, in: *Biorefinery of Alternative resources: Targeting green fuels and platform chemicals*, S. Nanda, D.V.N. Vo, P.K. Sarangi (Eds.), pp. 477–494, Springer Nature, Singapore, 2020, ISBN: 978-9811518034.

Bhatia, L., Sarangi, P.K., Nanda, S., Current Advancements in Microbial Fuel. *Cea. Eng. Technol.*, 43, 4, 601–612, 2020.

Faloye, F.D., Gueguim Kana, E.B., Schmidt, S., Optimization of biohydrogen inoculum development via a hybrid pH and microwave treatment technique-Semi-pilot scale production assessment. *Int. J. Hydrog. Energ.*, 39, 5607–5616, 2014.

Grimes, C.A., Varghese, O.K., Ranjan, S., *Light, Water, Hydrogen. The solar generation of hydrogen by water photoelectrolysis*, Springer Science+Business Media, LLC, New York, 2008, http://dx.doi.org/10.1007/978-0-387-68238-9.

Hallenbeck, P.C., *Microbial technologies in advanced biofuels production*, Springer Science+Business Media, LLC, New York, 2012.

Hawkes, F.R., Hussy, I., Kyazze, G., Dinsdale, R., Hawkes, D.L., Continuous dark fermentative hydrogen production by mesophilic microflora: principles and progress. *Int. J. Hydrog. Energy*, 32, 172–84, 2007.

Kadier, A., Simayi, Y., Kalil, M.S., Abdeshahian, P., Hamid, A.A., A review of the substrates used in microbial electrolysis cells (MECs) for producing sustainable and clean hydrogen gas. *Renew. Energy*, 71, 466e72, 2014.

Kotay, S.M. and Das, D., Biohydrogen as a renewable energy resource—prospects and potentials. *Int. J. Hydrog. Energy*, 33, 258–263, 2008.

Kumar, G., Bakonyi, P., Periyasamy, S., Kim, S.H., Nemestothy, N., Belafi-Bako, K., Lignocellulose biohydrogen: practical challenges and recent progress. *Renew Sustain. Energy Rev.*, 44, 728e37, 2015.

Lay, C., Sen, B., Lin, C.C.C., Continuous anaerobic hydrogen and methane production using water hyacinth feedstock. *Arabian J. Sci. Eng.*, 41, 7, 2563e71, 2016.

Min, B., Lim, J., Koa, S., Lee, K.G., Lee, S.H., Lee, S., Environmentally friendly preparation of pectins from agricultural byproducts and their structural/rheological characterization. *Bioresour. Technol.*, 102, 4, 3855–3860, 2011.

Nanda, S., N. Vo, DV., Sarangi, P., *Biorefinery of Alternative Resources: Targeting Green Fuels and Platform Chemicals*, Springer, Singapore, 2020.

Nanda, S., Rana, R., Vo, D.V.N., Sarangi, P.K., Nguyen, T.D., Dalai, A.K., Kozinski, J.A., A Spotlight on Butanol and Propanol as Next-Generation Synthetic Fuels, in: *Biorefinery of Alternative resources: Targeting green fuels and platform chemicals*, S. Nanda, D.V.N. Vo, P.K. Sarangi (Eds.), pp. 105–126, Springer Nature, Singapore, 2020, ISBN: 978-9811518034.

Reith, J., Wijffels, R., Barteb, H., Bio-methane & Bio-hydrogen: Status and perspectives of biological methane and hydrogen production, 2003, http://gasunie.eldoc.ub.rug.nl/root/2003/3339875/.

Renewables, Global status report, 2011, (www.ren21.net/status-of-renewables/global-status-report) Accessed September 25, 2015.

Rezania, S., Din, M.F.M., Taib, S.M., Sohaili, J., Chelliapan, S., Kamyab, H., *et al.*, Review on fermentative biohydrogen production from water hyacinth, wheat straw and rice straw with focus on recent perspectives. *Int. J. Hydrog. Energy*, 42, 20955e69, 2017.

Sarangi, P.K., Nanda, S., Vo, D.V.N., Recent Advances in Steam Reforming of Glycerol for Syngas Production, in: *Biorefinery of Alternative resources: Targeting green fuels and platform chemicals*, S. Nanda, D.V.N. Vo, P.K. Sarangi (Eds.), pp. 399–426, Springer Nature, 2020, ISBN: 978-9811518034.

Sarangi, P.K., Nanda, S., Sarangi, P.K., Nanda, S., Mohanty, P. (Eds.), Recent developments and challenges of acetone-butanol-ethanol fermentation, in: *Recent Advancements in Biofuels and Bioenergy Utilization*, pp. 111–123, Springer Nature, Singapore, 2018, ISBN No.: 978-981-13-1306-6.

Sarangi, P.K., Srivastava, R.K., Singh, A.K., Sahoo, U.K., Prus, P., Sass, R., Municipal-Based Biowaste Conversion for Developing and Promoting Renewable Energy in Smart Cities. *Sustainability*, 15, 12737, 2023, https://doi.org/10.3390/su151712737.

Sarangi, P.K., Subudhi, S., Bhatia, L., *et al.*, Utilization of agricultural waste biomass and recycling toward circular bioeconomy. *Environ. Sci. Pollut. Res.*, 30, 8526–8539, 2023, https://doi.org/10.1007/s11356-022-20669-1.

Sarangi, P.K., Vivekanand, V., Mohanakrishna, G., Pattnaik, B., Muddapur, U., Aminabhavi, T.M., Biorefinery of agricultural byproducts for the production of phenolic compounds: A bioeconomic perspective. *J. Clean. Prod.*, 414, 137460, 2023, https://doi.org/10.1016/j.jclepro.2023.137460.

Sarangi, P.K. and Nanda, S., Recent advances in consolidated bioprocessing for microbe-assisted biofuel production, in: *Fuel Processing and Energy Utilization*, S. Nanda, P.K. Sarangi, D.V.N. Vo (Eds.), pp. 141–157, CRC Press, Florida, USA, 2019, ISBN No.: 978-1-138-59320-6.

Sarangi, P.K. and Nanda, S., Bioconversion of agro-wastes into phenolic flavour compounds, in: *Biotechnology for Sustainable Energy and Products*, P.K. Sarangi and S. Nanda (Eds.), pp. 266–284, IK International Publishing House, India, 2019a.

Sarangi, P.K. and Nanda, S., Valorization of pineapple wastes for biomethane generation, in: *Biogas Technology*, S. Mishra, T.K. Adhya, S.K. Ojha (Eds.), pp. 169–180, New India Publishing Agency, New Delhi, India, 2019b, ISBN:978-93-87973-52-7.

Sarangi, P.K., Sustainable Utilization of Agro Wastes using Microbial Resources: A Future Perspective, in: *Microbe Assisted Sustainable Agriculture*, P.K. Sarangi and B. Sinha (Eds.), pp. 257–268, Overseas Press India Private Limited, New Delhi, India, 2019, ISBN: 978-81-939968-9-8.

Sarangi, P.K. and Nanda, S., Biohydrogen Production Through Dark Fermentation, in: *Chemical Engineering and Technology*, 2020.

Sarangi, P.K., Nanda, S., Vo, D.V.N., Technological Advancements in the Production and Application of Biomethanol, in: *Biorefinery of Alternative resources: Targeting green fuels and platform chemicals*, S. Nanda, D.V.N. Vo, P.K. Sarangi (Eds.), pp. 127–140, Springer Nature, Singapore, 2020, ISBN: 978-9811518034.

Sarangi, P.K., Singh, T.A., Singh Ng, J., Pineapple as potential crop resource: Perspective and value addition, in: *Food Bioresources and Ethnic Foods of Manipur*, pp. 83–91, Empyreal Publishing House, Northeast, India, 2019.

Show, K.Y., Lee, D.J., Tay, J.H., Lin, C.Y., Chang, J.S., Biohydrogen production: current perspectives and the way forward. *Int. J. Hydrog. Energ.*, 37, 20, 15616–15631, 2012.

Singh, A., Sevda, S., Abu Reesh, I., Vanbroekhoven, K., Rathore, D., Pant, D., *et al.*, Biohydrogen production from lignocellulosic biomass: technology and sustainability. *Energies*, 8, 13062e80, 2015.

Subudhi, S., Saha, K., Mudgil, D., Srivastava, R.K., Sarangi, P.K., Biomethanol production from renewable resources: a sustainable approach. *Environ. Sci. Pollut. Res.*, 32, 7432–7448, 2025, https://doi.org/10.1007/s11356-023-29616-0.

5

Biobutanol as Alternative Fuel: A Brief Outline on Current Advancements

Prakash Kumar Sarangi

College of Agriculture, Central Agricultural University, Imphal, Manipur, India

Abstract

Global energy demand is increasing vigorously due to urbanization and industrialization, thereby creating much dependence on conventional non-renewable fossil fuel sources. Automobile sectors are completely dependent on fossil-based sources, which create major environmental pollution. Hence, worldwide research is focused on seeking renewable energy-based fuels, which can provide valuable energy and materials. Among all fuels, biobutanol can be considered as a superior biofuel due to its higher energy content and lower hygroscopicity. Biobased feedstocks like waste biomass can be focused on for acetone-butanol-ethanol (ABE) fermentation. Converting renewable resources into valuable base chemicals and producing biobutanol gained a promising interest in the utilization of lignocellulosic biomass having no issues with food vs fuels. In this chapter, the various recent progress of ABE fermentation is discussed along with the major challenges and future possibilities.

***Keywords*:** Waste biomass, biobutanol, microorganism, acetone-butanol-ethanol fermentation, *Clostridium*

5.1 Introduction

Fossil-based sources are a major concern on energy demand. These not only produce greenhouse gases but also grow concern about global warming, the increase of the fossil price, and legislative restrictions for the use of non-renewable energy sources. These concerns forced an increase interest toward the exploration in biotechnology for biofuels production. However, the main global objective has been focused on the exploitation of

Prakash Kumar Sarangi. Waste Biomass to Alternative Fuels, (79–98)

renewable resources to produce fuels, power, heat, and value-added chemicals from renewable biomass. The drastic shift toward the increased interest for the replacement of fossil fuels is mainly due to global warming, the continued instability of oil prices, and the non-perennial availability of fossil fuels. Energy required for the chemical industry obtaining mainly from fuels such as petroleum and coal leads to the overall requirement toward renewable and eco-friendly technologies. Such advanced and well-known technologies that convert waste materials into various forms of secondary energy include electricity, gaseous and liquid biofuels, and even hydrogen. Lignocellulosic biomass (LCB) acting as a renewable source can produce future biofuels, mitigating the fossil fuel crisis. In general, there are various options to produce alternative transportation fuels based on biomass.

The expanding biofuel industry has faced important concerns although global biofuel production has been increasing rapidly over the last decade. Moreover, showing the sustainability of many first-generation biofuels has created many challenges over the food-vs-fuel debate, thereby exposing a better environment for the waste-to-energy scenario. Such a case in which LCB acts as the feedstocks for future biofuel production. Hence, biofuel technologies become efficient in terms of net lifecycle greenhouse gas (GHG) emission reductions, being socially and environmentally sustainable to reduce significant emission in the transport sector. Thus, the sustainability of second-generation feedstocks depend on whether producers comply with criteria like minimum lifecycle GHG reductions, including land use change and social standards.

The global consumption of petroleum and other fossil-based liquid fuels was 86 million barrels per day (BPD) in 2008, which was predicted for an increase to 98 million BPD in 2020 and 112 million BPD in 2035 (USEIA, 2011). Two important biofuels are used as the alternatives of fossil fuels, replacing gasoline in the future. There is great attention to the recovery of green energy in the form of liquid fuels from plant-based residues and other organic waste for developing countries.

Lignocellulosic feedstocks including agricultural and forestry residues, municipal solid wastes, and sewage sludge have the potential to supplement the production of such fuels, thereby achieving energy security and reducing GHG emissions (Nanda *et al.*, 2015b, 2014b). Being an inexpensive yet globally abundant resource, LCB can support the production of alternative liquid biofuels. About 40 million tons of LCB is generated annually (Sanderson, 2011). Other advantages of LCB lie in the nonedible nature, hence possessing a negligible threat to domestic and international food security unlike food-based fuel feedstocks (Nanda *et al.*, 2015b). Moreover, due to the noncompetition of LCB in food supply, they have a broad potential in bioethanol and biobutanol refineries. Although LCB is seen as the next generation of biofuels, there are many challenges in these aspects for the search of a superior biofuel due to some limitations of present biofuel. Various parameters like flexibility in raw material selection, better bioconversion, and more attractive fuel features than ethanol may be found toward the overcoming the limitations of present bioethanol.

In these aspects, butanol is regarded as a better competitor than ethanol due to many biofuel properties (Durre, 2007; Ni and Sun, 2009; Patakova *et al.*, 2013; Tigunova *et al.*, 2013; Karimi and Pandey, 2014; Li *et al.*, 2014). Moreover, butanol has gained many advantages over ethanol having 30% higher energy content and less corrosion, lower volatility, less flammability, and reduced hazardousness for handling (Nanda *et al.*, 2014a). Having low vapor pressure, butanol is used as biofuel in gasoline supply pipelines and well mixed with gasoline in flexible proportions (Qureshi and Ezeji, 2008). Butanol can be blended with gasoline in any ratio and can also be used as neat fuel in all gasoline engines. The biological production of butanol has been achieved through the acetone, butanol, and ethanol (ABE) fermentation process. Due to a high demand of acetone to produce cordite as explosive during the First World War, this process was commercialized in the Union of Soviet Socialist Republics, England, Canada, and the USA. The ABE fermentation process was developed in 1912 at Manchester University by Russian chemist C. Weizmann. Different classes of sugars and starches have been used to produce a mixture of ABE. Several reports have been found regarding establishments as industrial units in other countries including Japan, Australia, China, and South Africa (Linden *et al.*, 1986; García *et al.*, 2011; Köpke and Dürre, 2011; Dong *et al.*, 2012). Although the role of ABE fermentation was mainly to produce acetone as a solvent for military applications, there is an increasing interest in butanol as a liquid renewable fuel (Durre, 2007). Elevated substrate cost, low yield, and high cost of product recovery are major limitations in the economic production of butanol through ABE fermentation. The last two decades of global research explored alternate, cheap, renewable feedstock for ABE fermentation, such as bagasse, wheat straw, wheat bran, corn fiber, and other agricultural residue.

Having been bestowed various chemical and physical properties, butanol has been focused on producing second-generation biofuels. Butanol has four carbon atoms, 25% more energy with respect to ethanol, six times less evaporative than ethanol, 13.5 times less evaporative than gasoline, less hygroscopic than ethanol, thereby facilitated in existing transport systems of gasoline (Lee, 2008a). Butanol can be mixed at high ratios with gasoline for use in cars without modifications in gasoline-designed engines.

According to U.S. company Environmental Energy Inc., butanol can be used as a total replacement for gasoline without any modifications to car engines (Brekke, 2007). The important biofuel types of properties of biobutanol, gaining much interest toward global fuel industries are summarized in Table 5.1 with comparisons to other biofuels or alcohols.

Table 5.1 A comparison of properties of common biofuels with butanol.

Characteristics	Gasoline	Butanol	Ethanol	Methanol
Formula	H, C4 – C12	C_4H_9OH	CH_3CH_2OH	CH_3OH
Boiling Point (°C)	32 – 210	118	78	65
Auto-ignition temperature (°C)	280	343	365	435
Energy density (MJ kg –1)	44.5	33.1	26.9	19.6
Air fuel ratio	14.6	11.2	9.0	6.5
Research octane number	91 – 99	96	129	136
Viscosity at 25°C (mPa s)	0.6	2.573	1.074	0.544
Motor octane number	81 – 89	78	102	104
Density at 20°C (g/m3)	0.7	0.81	0.789	0.797
Lower heating value (MJ/kg)	43.4	34.3	26.9	22.7
Higher heating value (MJ/kg)	46.5	37.3	29.8	37.18
Heat of vaporization (MJ kg –1)	0.36	0.43	0.92	1.20

(Chan *et al.*, 2010; Cheng, 2010; Dürre, 2007; Lee *et al.*, 2008; Surisetty *et al.*, 2011; MacLean and Lave, 2003; Szulczyk, 2001)

Table 5.2 A list of biomasses and *Clostridium* spp. used for butanol production.

Bacteria	Feedstock	Total ABE concentration (g/L)	References
C. acetobutylicum P262	Aspen wood	20.1–24.6	Parekh *et al.*, 1988
C. acetobutylicum 824A	Lactose	1.43	Napoli *et al.*, 2010
C. acetobutylicum DSM 1731	Potato	33	Groben *et al.*, 1993
Clostridium acetobutylicum BKM19	Glucose	32.5	Jang *et al.*, 2013
C. saccharoperbutylacetonicum ATCC 27022	Bagasse	18.1	Soni *et al.*, 1982
C. beijerinckii P260	Barley straw	26.6	Qureshi *et al.*, 2010a
C. beijerinckii BA101	Corn fiber	9.3	Qureshi *et al.*, 2008b
C. acetobutylicum P262	Corn stover	25.8	Parekh *et al.*, 1988
C. acetobutylicum P262	Pinewood	17.6	Parekh *et al.*, 1988
C. beijerinckii B-592	Pinewood	18.5	Nanda *et al.*, 2014a
C. saccharoperbutylacetonicum ATCC 27022	Rice straw	13.0	Soni *et al.*, 1982
C. beijerinckii B-592	Wheat straw	17.9	Nanda *et al.*, 2014a
C. beijerinckii P260	Wheat straw	21.4	Qureshi *et al.*, 2008a
C. beijerinckii P260	Switchgrass	14.6	Qureshi *et al.*, 2010b

(*Continued*)

Table 5.2 A list of biomasses and *Clostridium* spp. used for butanol production. (*Continued*)

Bacteria	Feedstock	Total ABE concentration (g/L)	References
C. saccharobutylicum DSM 13864	Sago	9.1	Liew *et al.*, 2005
C. saccharoperbutylacetonicum ATCC 27022	Rice straw	13.0	Soni *et al.*, 1982
C. beijerinckii B-592	Timothy grass	17.4	Nanda *et al.*, 2014a

Several *Clostridium* spp, especially *Clostridium acetobutylicum* and *Clostridium beijerinckii*, have been utilized to produce biobutanol through the ABE fermentation process. Attempts have been made to obtain butanol from many starch-based and cellulosic substrates. Various fermentation technologies like batch, fed-batch, and continuous fermentation using native and modified strains are applied in the ABE fermentation process (Bankar *et al.*, 2012; Setlhaku *et al.*, 2012; Survase *et al.*, 2012; Xue *et al.*, 2012; Chen *et al.*, 2013; Chen *et al.*, 2014). A list of *clostridium* spp along with feedstocks used in ABE fermentation process is summarized in Table 5.2.

5.2 The ABE Fermentation Process

Clostridium bacterium is a heterogeneous, gram-positive, spore-forming, naturally rod-shaped obligate anaerobe that can grow on a wide range of substrates, producing many enzymes and breaking complex polymeric carbohydrates into monomers. However, the catalytic effect is drawn by various types of enzymes such as α-amylase. α-glycosidase, glucoamylase, pullulanase, and amylopullulanase (Ezeji *et al.*, 2007a). Availability of various suitable biocatalytic systems in *Clostridium* can enhance the production of butanol over broad ranges of carbon sources (e.g. arabinose, cellobiose, galactose, glucose, mannose, and xylose) to ABE (Ezeji *et al.*, 2007b). The biological production of butanol is only achieved by the ABE fermentation process, which has long been a widely accepted phenomenon (Figure 5.1). Butanol production by ABE fermentation is mostly carried out using *Clostridium* spp. like *C. acetobutylicum* and *C. Beijerinckii*. Other

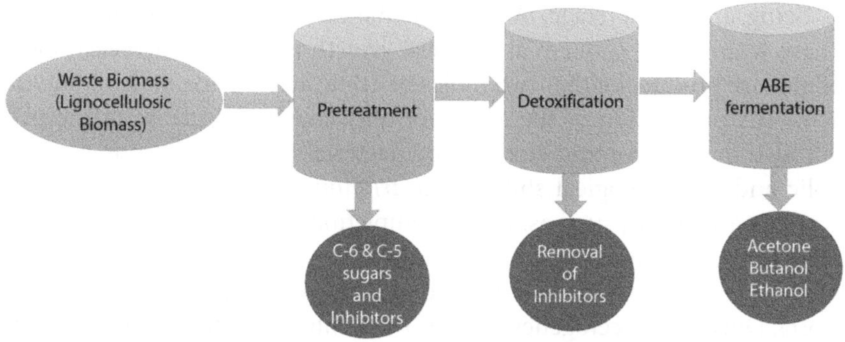

Figure 5.1 An outline of steps for biobutanol production through ABE fermentation.

species including *C. acetobutylicum*, *C. beijerinckii*, *C. aurantibutyricum*, *C. butylicum*, *C. saccharobutylicum*, *C. saccharoperbutylacetonicum*, can also be used in ABE fermentation with a variety of organic substrates. Some traditional starch-based feedstocks for butanol production include corn, millet, molasses, potatoes, rice, wheat, whey permeate, and tapioca. These restrict the requirement of biomass pretreatment and hydrolysis (Qureshi and Blascheck, 2005) due to the intense amylolytic activity of *Clostridium*. Over the first generation of fuels, a focus toward the production of butanol from LCB is attained due to the concerns of the food-vs-fuel scenario. Various LCB including crop residues and agro-waste, such as barley straw, corn cobs, corn fiber, corn stover, hemp waste, pinewood, rice straw, sunflower shells, switchgrass, Timothy grass, and wheat straw are used for biobutanol production through ABE fermentation (Zverlov, 2006; Qureshi and Ezeji, 2008; Nanda *et al.*, 2014).

Having a complex intracellular pathway, ABE fermentation principally produces three types of important products such as (1) solvents (acetone, butanol, and ethanol); (2) organic acids (lactic acid, acetic acid, and butyric acid); and (3) gases (carbon dioxide and hydrogen) (Zheng *et al.*, 2009; Xue *et al.*, 2013). The entire ABE fermentation is a biophasic process that consists of both the acidogenic phase and the solventogenic phase. In a typical batch system, ABE fermentation is initiated with carbohydrate substrates (usually glucose or equivalent sugars). The fermentation medium is created by an anaerobic environment inside the reactor using N_2 or CO_2 followed by a suitable, butanol-producing *Clostridium* culture incubating at 35°C. The typical incubation period of ABE fermentation lasts for 36–72 h, producing total ABE up to 20–25 g/L (Qureshi and Ezeji, 2008).

Being a biphasic stage, ABE fermentation starts with the acidogenic phase within the exponential bacterial growth phase in which each mole of glucose produces either two moles of acetic acid or one mole of butyric acid. The production of the acids not only causes a decrease in the pH, but also favors no more suitable for bacterial growth. There is a metabolic and morphological shift by *Clostridium acetobutylicum* during the acetogenesis phase, such as: i) The completion of the exponential growth phase; ii) Endospores formation, cessation of growth; and iii) Conversion of acids to solvents, acetone-butanol-ethanol in a typical molar ratio of 3:6:1). During the acetogenesis phase, the inhibition of the metabolic pathway can also occur even after no proper pH control, which is regarded as acidic stress. However, the causes of acidic stress is the rapid production of these acids compared to their consumption (Kumar and Gayen, 2011; Xue *et al.*, 2013). Another aspect of this phase is that exponential growth stage of *Clostridium* favors the acidogenic phase as the formation of acid is complemented by the synthesis of adenosine triphosphate required for cell growth. Being an obligate anaerobe, the acidogenic phase plays a vital role in energy metabolism of *Clostridium.* Due to the reduction of pH, bacterium slows down acid production and utilizes the excreted acetic acid and butyric acid, thereby converting them to acetone and butanol, respectively. During the acidogenic phase, more butyric acid is produced than acetic acid because of a better balance of the redox equilibrium of the former (Zheng, 2009). The formation of butyric acid needs nicotinamide adenine dinucleotide hydrogen (NADH) from glycolysis rather than acetate formation, thereby producing more butanol from butyric acid than ethanol. Hence, more conversion of butyrate than acetate occurs during the solventogenic phase in ABE fermentation, almost twice as much as butyrate produced compared to acetate (Dürre, 2007). Thus, butanol is the most important product of ABE fermentation as the typical proportions of acetone, butanol, and ethanol are 3:6:1 (Jones and Woods, 1986).

Clostridium metabolism involves the phase shift from acidogenesis to solventogenesis, thereby products of the former are then transferred to the latter followed by a 70%–80% conversion of viable cells into spores, leading to a cessation of growth (García *et al.*, 2011). At around pH 5.5, the acid concentration remains nearly constant or slightly lower, thereby facilitating the production of the solvent (ABE). The causative parameters of a phase shift may be due to non-dissociated butyric acid (García *et al.*, 2011), as well as with butyric acid, acetic acid also induces respective enzymes such as butyrate kinase and acetate kinase needed for the butanol biosynthesis (Ballongue *et al.*, 1986). Moreover, the role of butyryl-CoA and butyrylphosphate is also there to shift from the aectogenesis phase to the solventogenesis phase (Harris *et al.*, 2000). Acetic acid can be converted to ethanol or acetone, while butyric acid is converted to butanol (Figure 5.2). The cessation of the fermentation process becomes of high concentration of the solvents, inhibiting the process so that cell membranes are solubilized ultimately to cell death. However, the maximum solvent concentration of about 2 wt% is the limitation of this ABE fermentation process (Durre, 1998).

The anaerobic ABE fermentation, lasting for 36–72 h produces total ABE of up to 20–25 g/L (Qureshi and Ezeji, 2008). At the solventogenesis phage, the accumulated solvents become toxic to the cells, assuming 50% of the cell growth is stalled by concentrations of 11, 51, and 84 g/l butanol, ethanol, and acetone, respectively. However, butanol is very toxic to the cells showing inhibition at a concentration range of 5–10 g/L because *Clostridia* can rarely withstand more than 2% butanol (Qureshi and Ezeji, 2008; Liu and Qureshi, 2009). There are various research cases available on solvent stress (Linden *et al.*, 1986; Xue *et al.*, 2013) due to the sensitivity of *Clostridia* to the medium composition and fermentation conditions. Even small amounts of oxygen can completely inhibit the activity of the cells, and some quantities of chemicals can also affect product distribution (Choi *et al.*, 2012; Han *et al.*, 2013) by a case study of a small amount of zinc, e.g., 0.001 g/L $ZnSO4.7H_2O$ that can result in an earlier shifting to solventogenesis (Wu *et al.*, 2013).

The role of pH in fermentation conditions is very important for determining the production of acids and solvents. Basically, during the

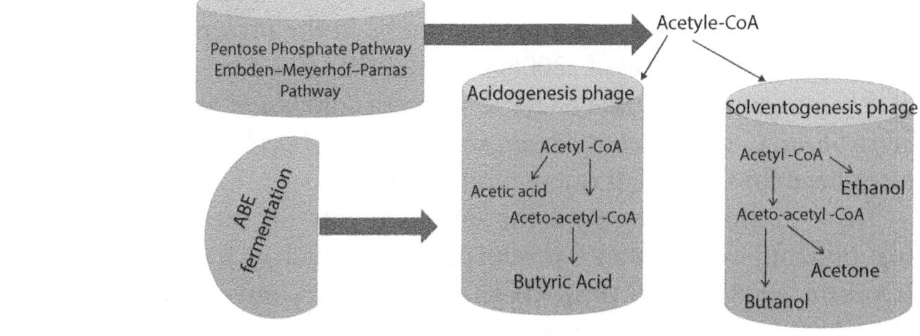

Figure 5.2 Simplified pathways for ABE fermentation by *Clostridium*.

acetogenesis phage, the formation of acetate and butyrate lowered the pH level, thereby facilitating towards the solventogenic phage. At a crucial pH level, the solventogenic phase starts assimilating the produced acid into acetone and butanol (Lee *et al.*, 2008). Although low pH is preferable for solvent production, a pH below 4.5 results in enough acid formation, thus lowering the duration and effectiveness of the solventogenic phase. However, the enhancement of bacterial growth can occur by increasing the buffering capacity of the fermentation medium so that carbohydrate utilization and butanol production can also be increased (Bryant and Blaschek, 1988).

The best studied bacterium is the *C. acetobutylicum* that basically grows on starch-based substrates. Some other phylogenetically interconnected strains are *C. beijerinckii*, *C. saccharobutylicum*, and *C. saccharoperbutylacetonicum* having saccharolytic activities. The production of solvents at concentrations up to 14–18 g L–1 with solvent yields of 25%–30% (Shaheen *et al.*, 2000) by some the strains of *C. acetobutylicum*. The production of solvents over a wide range of pH occurs by the saccharolytic industrial strains belonging to the *C. beijerinckii* species, utilizing a wider variety of carbohydrates as a substrate due to its genetic potential. *C. beijerinckii* is also less susceptible to acid crash. Therefore, it is more suitable for longer (continuous) fermentations than *C. acetobutylicum* (Qureshi, and Blaschek, 2000). The less studied clostridium are the *C. saccharobutylicum* and *C. saccharoperbutylacetonicum*, producing average solvent concentration of 19.6g L–1 with a yield of 30% (Shaheen *et al.*, 2000).

The role of electron flow in the glycolytic pathway is very crucial to produce butanol from the ABE fermentation process. The presence of ferredoxin oxidoreductase in *Clostridia* helps in the oxidation of NADH and flavin adenine dinucleotide hydrogen (FADH) produced during the solventogenic phase. Any alteration in the direction of electron flow around reduced ferredoxin can alter the type and quantity of fermentation products. The enhancing effect on the production of butanol and ethanol is detected at the expense of acetone synthesis in the presence of these electron carriers (Zverlov *et al.*, 2006; Gapes, 2000). The four principal groups of solventogenic *Clostridia* bacteria such as *Clostridium acetobutylicum*, *C. beijerinckii*, *C. saccharobutylicum*, and *C. Saccharoperbutylacetonicum* may be utilized for the ABE fermentation process.

There is attention toward the developments on metabolic engineering, utilizing hyper butanol-producing strains such as *C. beijerinckii* P260 (Qureshi 2008) and *C. beijerinckii* BA101. Also, novel separation technologies can prevent the product inhibition process due to the removal of accumulated butanol from the fermentation medium (Lee *et al.*, 2008).

As compared to the final solvents accumulation 12 to 20 g/L (Lee *et al.*, 2008) in batch fermentation, the fed-batch fermentation has more product recovery but is facing more problems on butanol toxicity, which can be reduced by the supplementation of *in situ* recovery processes such as gas stripping (Eteji *et al.*, 2004).

5.3 Challenges and Possible Outcomes in the ABE Fermentation

Having many advantages over ethanol, the commercial production of butanol through ABE fermentation has great challenges and bottlenecks at various stages of process. Although LCB has been used as cheap sources of substrates for ABE fermentation, the main challenge of using LCB as feedstock is the additional cost of sugar production from lignocellulose, compared to other feed stock like molasses or starches. Another limitation is the build-up at the selection of biomass along with the pretreatment processes. Whatever the pretreatment process, detoxification is crucial for the removal of inhibitors generated during these processes (Ezeji *et al.*, 2007). Apart from other challenging technical issues associated with ABE fermentation common to all feedstocks, butanol toxicity and low recovery can hinder its commercial production, which significantly increases the cost of recovery and separation (Ezeji *et al.*, 2007). Hence, the production cost of biobutanol becomes more expensive (Qureshi *et al.*, 2008c), selling prices per liter to be higher than that of gasoline, thereby uncompetitive in the transportation fuels. There is a need for global extensive research in the process development and bioconversion technology models to reduce costs of biobutanol production. These are a few challenges and possible outcomes of ABE fermentation to produce butanol toward their mitigation, thereby exposing commercial butanol for solving a fuel crisis in future.

Although interest in the sustainable production of butanol from renewable biomass is increasing among academicians and industrial sectors (Jung *et al.*, 2013; Gao *et al.*, 2014), the cost of the substrate only accounts for 60% of the overall production cost. Hence, low cost and year-round availability are the key issues for the success of biofuel production by the biotechnological route. Different pretreatment methods such as physical, chemical, and biological techniques have been well investigated for biofuel production from LCB (Kumar *et al.*, 2009; Mussatto and Teixeira, 2010), aiming to increase the accessibility of cellulose and hemicelluloses for obtaining fermentable monomeric sugars (Galbe and Zacchi, 2007).

The ABE fermentation process can be successful and commercially exposed in the global market by applying only suitable biomass, which is inexpensive and abundant by nature (Lepiz–Aguilar *et al.*, 2013; Xue *et al.*, 2013; Zhang *et al.*, 2014; Becerra *et al.*, 2015; Kheyrandish *et al.*, 2015). Lignocellulosic biomass are regarded as suitable substrates for conversion into biobutanol through ABE fermentation, but because of recalcitrants and the inexpensive nature of hydrolytic enzymes, the price of butanol would have increased many folds, thereby limiting its commercial production (Shafiei *et al.*, 2011; Shafiei *et al.*, 2013; Shafiei *et al.*, 2014, Boonsombuti *et al.*, 2015). Compared to ethanol production by yeasts, fermentation through ABE-processed microorganisms have to possibly face some inhibitors present in lignocelluloses hydrolysates, e.g., hydroxymethylfurfural, furfural, and lignin derivatives (Kudahettige–Nilsson *et al.*, 2015). Not only are there inhibitors that slow the growth of the Clostridium species but the yield of butanol is also reduced (Cai *et al.*, 2013). Hence, a suitable pretreatment process has to be considered before the initiation of ABE fermentation on the account formation of various inhibitors. An ideal pretreatment process should efficiently improve the enzymatic hydrolysis, consume lower amounts of chemicals, and produce fewer by-products/inhibitors (Karimi *et al.*, 2013). Compared to ethanol fermentation yeasts. For instance, dilute acid pretreatment, which produces a high concentration of inhibitors, should not be applied. Otherwise, an extra detoxification process that also consumes chemicals is necessary and is accompanied by some sugar loss and a production of problematic wastes. Liquid hot water, ammonia, ionic liquid, and organosolv treatments are among the most applied methods (Amiri *et al.*, 2014), but all these methods have their own drawbacks (Taherzadeh and Karimi, 2008).

The major limitation of the industrial production of ABE is butanol toxicity. Fermentation by *Clostridium* spp has to face such hurdles as said microbes rarely tolerate more than 2% butanol. Butanol toxicity is a limiting factor for the final butanol yield in ABE fermentation, which is more severe than acetone and ethanol. The maximum limit tolerated by wild-type *Clostridium* strains (Garcia 2011) is reported to be butanol level up to 12–13 g/L by the conventional ABE fermentation. Butanol concentration of 19.6 g/L is reportedly produced from genetically modified *C. beijerinckii BA101* and is found to be at maximum limit (Qureshi and Blaschek, 1999). Another researchable issue that hinders the butanol production level is the bacteriophage infection during ABE fermentation. Some report states the Siphoviridae and Podoviridae infect *Clostridium madisonii* and *C. beijerinckii P260*, respectively, thereby lowering the growth of bacterium, thus showing reduced solvent production (Jones, 2000). Although butanol has partial miscibility in water, this feature leads to several technical issues during its recovery from the fermentation broth. Starting from distillation as a traditional solvent recovery method, various techniques such as perstraction (Qureshi, 1992), pervaporation (Xue, 2014) and supercritical methods (Reddy, 2014) have been undertaken for maximum butanol recovery. The recent development towards genetic and metabolic engineering has also applied for butanol-producing microorganisms to overcome several limitations, such as low butanol titer, yield, and productivity. The construction of mutant strains of *Clostridium* and other microorganisms are the vital factors for improvement of industrial scale butanol refinery (Eteji, 2007). Another new technology for enhancing microbial efficiency for butanol production is antisense RNA technology implemented in producing mutants for improved ABE fermentation (Tummala, 2003). Hence, a synthetic biological approach can be implemented to develop better microbial strains tolerant to a high butanol concentration and other solvent stress. This can mitigate the future fuel crisis due to the decline and adverse effects of fossil fuel uses.

5.4 Conclusion

Biobutanol is an assured substitute of other biofuels and petroleum-based products. The major merits of butanol are the high energy content and less corrosion than ethanol, being a pure alcohol with a longer hydrocarbon chain, thereby being directly used as fuel in existing internal combustion engines in either blended or pure form. The ABE fermentation process has already been utilized to produce acetone, butanol, and ethanol in the ratio of 3:6:1 utilizing *Clostridium* species along with acetic acid, butyric acid, CO_2, and H_2 as gaseous products of ABE fermentation. During ABE fermentation, *Clostridium* is prone to butanol toxicity, spore formation, opportunistic bacteriophage infection and incomplete sugar conversion.

Nevertheless, this process has gained renewed interest due to its unique merits. Most ABE fermentation processes deal with the laboratory scale restricted to research based on a huge literature study. Hence, the pilot scale production of butanol utilizing the ABE fermentation process is a great challenge, taking the efforts of microbiologists, biotechnologists, and chemical engineers globally to produce a great substitute for gasoline. Comprising of two types of phases like acidogenesis and solventogenesis, the ABE fermentation process has been facing several challenges, such as low concentration, difficulty in the separation of products, higher feedstock consumption rate (i.e., lower production yield), and sensitivity to substrate composition, inhibitors, and to the presence of oxygen (Karimi *et al.*, 2015).

Metabolic engineering of butanol-producing microorganisms can induce oxygen and butanol tolerance, high cell density, prolonged viability, sporogenesis, and high butanol selectivity. Thus, the development of genetically modified and metabolically engineered strains can produce higher quantities of butanol, lowering side chain products that create a challenge set unit for scientific and academic communities around the world (Blaschek, 1993; Woods, 1995; Blaschek and White, 1995; Girbal and Soucaille, 1998). Some of the challenges like the high price of hydrolytic enzymes and expensive pretreatment processes are should be overcome to deal with LSB waste as promising feedstocks for ABE production in the future. More research work by academicians and researchers around the world must refurbish old ABE production processes toward the efficient production of second-generation butanol in the future economically and ecologically. Due to the unique physiology of *Clostridium* and a lack of understanding of its induced genomic regulation, recombinant DNA technology still struggles to develop a hyper butanol-producing strain. An improvement in biotechnological processes, bioprocess technology, and a consolidated bioprocessing system involving ABE fermentation may solve some of the challenges and difficulties for higher butanol production that could make sustainable fuel replacement for gasoline in future.

References

Amiri, H., Karimi, K., Zilouei, H., Organosolvent pretreatment of rice straw for efficient acetone, butanol, and ethanol production. *Bioresour. Technol.*, 152, 450–456, 2014.

Ballongue, J., Amine, J., Masion, E., Petitdemange, H., Gay, D., Regulation of acetate kinase and butyrate kinase by acids in *Clostridium acetobutylicum*. *FEMS Microbiol. Lett.*, 35, 295–301, 1986.

Becerra, M., Cerdan, M.E., Gonzalez-Siso, M.I., Biobutanol from cheese whey. *Microb. Cell Fact.*, 14, 27, 2015.

Blaschek, H.P., Recent developments in the genetic manipulation of microorganisms for biotechnology applications. *Food Process*, 2, 459–474, 1993.

Blaschek, H.P. and White, B.A., Genetic systems development in the clostridia. *FEMS Microbiol. Rev.*, 17, 349–356, 1995.

Boonsombuti, A., Luengnaruemitchai, A., Wongkasemjit, S., Effect of phosphoric acid pretreatment of corncobs on the fermentability of *Clostridium beijerinckii* TISTR 1461 for biobutanol production. *Prep. Biochem. Biotechnol.*, 45, 173–191, 2015.

Brekke, K., Butanol: An Energy Alternative?, in: *Ethanol Today*, pp. 36–39, 2007.
Bryant, D.L. and Blaschek, H.P., Buffering as a means for increasing growth and butanol production by *Clostridium acetobutylicum. J. Ind. Microbiol.*, 3, 49–55, 1988.
Cai, D., Zhang, T., Zheng, J., Chang, Z., Wang, Z., Qin, P.Y., Tan, T.W., Biobutanol from sweet sorghum bagasse hydrolysate by a hybrid pervaporation process. *Bioresour. Technol.*, 145, 97–102, 2013.
Chan, C.C., Horng, J.L., Chi, M.S., Yen, C.H., Autoignition Temperature Data for Methanol, Ethanol, Propanol, 2-Butanol, 1-Butanol, and 2-Methyl-2,4-pentanediol. *J. Chem. Eng. Data*, 55, 11, 5059–5064, 2010.
Chen, C., Wang, L., Xiao, G., Liu, Y., Xiao, Z., Deng, Q., Yao, P., Continuous acetone-butanol-ethanol (ABE) fermentation and gas production under slight pressure in a membrane bioreactor. *Bioresour. Technol.*, 163, 6–11, 2014.
Chen, C., Xiao, Z., Tang, X., Cui, H., Zhang, J., Li, W., Ying, C., Acetone-butanol-ethanol fermentation in a continuous and closed-circulating fermentation system with PDMS membrane bioreactor. *Bioresour. Technol.*, 128, 246–251, 2013.
Cheng, *Biomass to Renewable Energy Processes*, 1st, vol. 8, p. 272, CRC Press, New York, 2010.
Choi, S.J., Lee, J., Jang, Y.S., Park, J.H., Lee, S.Y., Kim, I.H., Effects of nutritional enrichment on the production of acetone-butanol-ethanol (ABE) by *Clostridium acetobutylicum. J. Microbiol.*, 50, 063–1066, 2012.
Durre, P., New insights and novel developments in clostridial acetone/butanol/isopropanol fermentation. *Appl. Microbiol. Biotechnol.*, 49, 639–648, 1998.
Dürre, P., Biobutanol: an attractive biofuel. *Biotechnol. J.*, 2, 1525–34, 2007.
Dong, H., Tao, W., Dai, Z., Yang, L., Gong, F., Zhang, Y., Li, Y., Biobutanol. *Adv. Biochem. Eng./Biotechnol.*, 128, 85–100, 2012.
Ezeji, T.C., Qureshi, N., Blaschek, H.P., Bioproduction of butanol from biomass: from genes to bioreactors. *Curr. Opin. Biotechnol.*, 18, 220–7, 2007.
Ezeji, T.C., Qureshi, N., Blaschek, H.P., Acetone butanol ethanol (ABE) production from concentrated substrate: reduction in substrate inhibition by fed-batch technique and product inhibition by gas stripping. *Appl. Microbiol. Biotechnol.*, 63, 653–8, 2004.
Ezeji, T., Qureshi, N., Blaschek, H., Butanol production from agricultural residues: impact of degradation products on *Clostridium beijerinckii* growth and butanol fermentation. *Biotechnol. Bioeng.*, 97, 1460–1469, 2007b.
Gao, K., Rehmann, L., Boiano, S., Marzocchella, A., Cellulosic Butanol Production from Alkali-PretreatedSwitchgrass (Panicumvirgatum) and Phragmites (Phragmitesaustralis). *Bioresour. Technol.*, 174, 176–181, 2014.
Galbe, M. and Zacchi, G., Pretreatment of lignocellulosic materials for efficient bioethanol production. *Adv. Biochem. Engin. Biotechnol.*, 108, 41–65, 2007.

García, V., Päkkilä, J., Ojamo, H., Muurinen, E., Keiski, R.L., Challenges in biobutanol production: How to improve the efficiency? *Renew. Sustain. Energy Rev.*, 15, 964–980, 2011.

Gapes, J., The Economics of Acetone-Butanol Fermentation: Theoretical and Market Considerations. *J. Mol. Microbiol. Biotechnol.*, 2, 27–32, 2000.

Girbal, L. and Soucaille, P., Regulation of solvent production in *Clostridium acetobutylicum*. *Trends Biotechnol.*, 16, 11–16, 1998.

Grobben, N.G., Eggink, G., Cuperus, F.P., Huizing, H.J., Production of acetone, butanol and ethanol (ABE) from potato wastes: fermentation with integrated membrane extraction. *Appl. Microbiol. Biotechnol.*, 39, 494–8, 1993.

Harris, L.M., Desai, R.P., Welker, N.E., Papoutsakis, E.T., Characterization of recombinant strains of the *Clostridium acetobutylicum* butyrate kinase inactivation mutant: need for new phenomenological models for solventogenesis and butanol inhibition? *Biotechnol. Bioeng.*, 67, 1–11, 2006.

Han, B., Ujor, V., Lai, L.B., Gopalan, V., Ezeji, T.C., Use of proteomic analysis to elucidate the role of calcium in acetone-butanol-ethanol fermentation by *Clostridium beijerinckii* NCIMB 8052. *Appl. Environ. Microbiol.*, 79, 282–293, 2013.

Jang, Y.S., Malaviya, A., Lee, S.Y., Acetone-butanol-ethanol production with high productivity using *Clostridium acetobutylicum* BKM19. *Biotechnol. Bioeng.*, 110, 6, 1646–53, Jun. 2013.

Jones, D.T. and Woods, D.R., Acetone-butanol fermentation revisited. *Microbiol Rev.*, 50, 4, 484–524, 1986. doi: 10.1128/mr.50.4.484-524.1986

Jung, Y.H., Kim, I.J., Kim, H.K., Kim, K.H., Dilute acid pretreatment of lignocellulose for whole slurry ethanol fermentation. *Bioresour. Technol.*, 132, 109–14, 2013.

Jones, D.T., Shirley, M., Wu, X., Keis, S., Bacteriophage infections in the industrial acetone butanol (AB) fermentation process. *J. Mol. Microbiol. Biotechnol.*, 2, 21–6, 2000.

Karimi, K., Shafiei, M., Kumar, R., Progress in physical and chemical pretreatment of lignocellulosic biomass, in: *Biofuel Technologies*, V.K. Gupta, and M.G. Tuohy (Eds.), pp. 53–96, Springer Berlin Heidelberg, Berlin, 2013.

Karimi, K. and Pandey, A., Current and future ABE processes. *Biofuel Res. J.*, 3, 77, 2014.

Karimi, K., Tabatabaei, M., Sárvári, H.I., Kumar, R., Recent trends in acetone, butanol, and ethanol (ABE) production. *Biofuel Res. J.*, 8, 301–308, 2015.

Kheyrandish, M., Asadollahi, M.A., Jeihanipour, D.M., Rismani-Yazdi, H., Karimi, K., Direct production of acetone–butanol–ethanol from waste starch by free and immobilized *Clostridium acetobutylicum*. *Fuel*, 142, 129–133, 2015.

Köpke, M. and Dürre, P., Biochemical production of biobutanol, in: *Handbook of Biofuels Production*, Woodhead Publishing, pp. 221–257, 2011.

Kudahettige-Nilsson, R.L., Helmerius, J., Nilsson, R.T., Sjoblom, M., Hodge, D.B., Rova, U., Biobutanol production by *Clostridium acetobutylicum* using xylose recovered from birch Kraft black liquor. *Bioresour. Technol.*, 176, 71–79, 2015.

Kumar, P., Barrett, D.M., Delwiche, M.J., Stroeve, P., Methods for pretreatment of lignocellulosic biomass for efficient hydrolysis and biofuel production. *Ind. Eng. Chem. Res.*, 48, 8, 3713–3729, 2009.

Kumar, M. and Gayen, K., Developments in biobutanol production: New insights. *Appl. Energy*, 88, 1999–2012, 2011.

Lee, S.Y., Park, J.H., Jang, S.H., Nielsen, L.K., Kim, J., Jung, K.S., Fermentative butanol production by *Clostridia*. *Biotechnol. Bioeng.*, 101, 209–28, 2008.

Lepiz-Aguilar, L., Rodriguez-Rodriguez, C.E., Arias, M.L., Lutz, G., Acetone-Butanol-Ethanol (ABE) production in fermentation of enzymatically hydrolyzed cassava flour by *Clostridium beijerinckii* BA101 and solvent separation. *J. Microbiol. Biotechnol.*, 23, 1092–1098, 2013.

Liew, S.T., Arbakariya, A., Rosfarizan, M., Raha, A.R., Production of solvent (acetone–butanol–ethanol) in continuous fermentation by *Clostridium saccharobutylicum* DSM 13864 using gelatinised Sago starch as a carbon source. *Malays J. Microbiol.*, 2, 42–5, 2005.

Li, J., Baral, N.R., Jha, A.K., Acetone-butanol-ethanol fermentation of corn stover by Clostridium species: present status and future perspectives. *World J. Microbiol. Biotechnol.*, 30, 1145–1157, 2014.

Linden, J.C., Moreira, A.R., Lenz, T.G., Acetone and butanol, in: *Comprehensive biotechnologyEngland*, M. Moo Young, (Ed.), pp. 915–931, Pergamon Press Oxford, 1986.

Liu, S. and Qureshi, N., How microbes tolerate ethanol and butanol? *New Biotechnol.*, 26, 117–21, 2009.

MacLean, H.L. and Lave, L.B., Evaluating automobile fuel/propulsion system technologies. *Prog. Energy Combust. Sci.*, 29, 1–69, 2003.

Mussatto, S.I. and Teixeira, J.A., Lignocellulose as raw material in fermentation processes. *Appl. Microb.*, 897–907, 2010.

Nanda, S., Mohammad, J., Reddy, S.N., Kozinski, J.A., Dalai, A.K., Pathways of lignocel-lulosic biomass conversion to renewable fuels. *Biomass Conv. Bioref.*, 4, 157–191, 2014b.

Nanda, S., Azargohar, R., Dalai, A.K., Kozinski, J.A., An assessment on the sustainability of lignocellulosic biomass for biorefining. *Renew. Sust. Energ. Rev.*, 50, 925–941, 2015b.

Nanda, S., Dalai, A.K., Kozinski, J.A., Butanol and ethanol production from lignocellulosic feedstock: Biomass pretreatment and bioconversion. *Energ. Sci. Eng.*, 2, 138–148, 2014a.

Napoli, F., Olivieri, G., Russo, M.E., Marzocchella, A., Salatino, P., Production of butanol in a continuous packed bed reactor of *Clostridium acetobutylicum*. *J. Ind. Microbiol. Biotechnol.*, 37, 603–8, 2010.

Ni, Y. and Sun, Z., Recent progress on industrial fermentative production of acetone-butanol-ethanol by *Clostridium acetobutylicum* in China. *Appl. Microbiol. Biotechnol.*, 83, 415–423, 2009.

Parekh, S.R., Parekh, R.S., Wayman, M., Ethanol and butanol production by fermentation of enzymatically saccharified SO2-prehydrolysed lignocellulosics. *Enzyme Microbial. Technol.*, 10, 660–668, 1988.

Patakova, P., Linhova, M., Rychtera, M., Paulova, L., Melzoch, K., Novel and neglected issues of acetone-butanol-ethanol (ABE) fermentation by clostridia: Clostridium metabolic diversity, tools for process mapping and continuous fermentation systems. *Biotechnol. Adv.*, 31, 58–67, 2013.

Qureshi, N. and Blaschek, H.P., Production of acetone butanol ethanol (ABE) by a hyper-producing mutant strain of *Clostridium beijerinckii* BA101 and recovery by pervaporation. *Biotechnol. Prog.*, 15, 594–602, 1999.

Qureshi, N., Maddox, I.S., Friedl, A., Application of continuous substrate feeding to the ABE fermentation: relief of product inhibition using extraction, perstraction, stripping and pervaporation. *Biotechnol. Prog.*, 8, 382–90, 1992.

Qureshi, N. and Blascheck, H.P., Butanol production from agricultural biomass, in: *Food biotechnology*, A. Pometto, K. Shetty, G. Paliyath, R.E. Levin (Eds.), pp. 525–49, CRC Press, Boca Raton, 2005.

Qureshi, N. and Ezeji, T.C., Butanol, 'a superior biofuel' production from agricultural residues (renewable biomass): recent progress in technology. *Biofuel Bioprod. Biorefin.*, 2, 319–30, 2008.

Qureshi, N. and Blaschek, H., Butanol Production using *Clostridium beijerinckii* BA101 and recovery by pervaporation. *Appl. Biochem. Biotechnol.*, 84, 225–230, 2000.

Qureshi, N., Saha, B.C., Cotta, M.A., Butanol production from wheat straw hydrolysate using *Clostridium beijerinckii*. *Bioprocess. Biosyst. Eng.*, 30, 419–27, 2007.

Qureshi, N., Saha, B.C., Dien, B., Hector, R.E., Cotta, M.A., Production of butanol (a biofuel) from agricultural residues: Part I—Use of barley straw hydrolysate. *Biomass Bioenerg.*, 34, 559–565, 2010a.

Qureshi, N., Ezeji, T.C., Ebener, J., Dien, B.S., Cotta, M.A., Blaschek, H.P., Butanol production by *Clostridium beijerinckii*. Part I: Use of acid and enzyme hydrolyzed corn fiber. *Bioresour. Technol.*, 99, 5915–5922, 2008b.

Qureshi, N., Saha, B.C., Hector, R.E., Dien, B., Hughes, S., Liu, S., Iten, L., Sarath, G., Cotta, M.A., Production of butanol (a biofuel) from agricultural residues: Part II - Use of corn stover and switchgrass hydrolysates. *Biomass Bioenerg.*, 34, 566–571, 2010b.

Qureshi, N., Saha, B.C., Hector, R.E., Hughes, S.R., Cotta, M.A., Butanol production from wheat straw by simultaneous saccharification and fermentation using *Clostridium beijerinckii*: Part I—Batch fermentation. *Biomass Bioenerg.*, 32, 168–75, 2008a.

Qureshi, N. and Ezeji, T., Butanol, 'a superior biofuel' production from agricultural residues (renewable biomass). *Rec. Prog. Techn.*, 2, 4, 319–330, 2008.

Reddy, S.N., Nanda, S., Dalai, A.K., Kozinski, J.A., Supercritical water gasification of biomass for hydrogen production. *Int. J. Hydrogen Energy*, 39, 6912–26, 2014.

Sanderson, K., Lignocellulose: A chewy problem. *Nature*, 474, S12–S14, 2011.

Setlhaku, M., Brunberg, S., Villa Edel, A., Wichmann, R., Improvement in the bioreactor specific productivity by coupling continuous reactor with repeated fed-batch reactor for acetone-butanol-ethanol production. *J. Biotechnol.*, 161, 147–152, 2012.

Shafiei, M., Karimi, K., Taherzadeh, M.J., Techno-economical study of ethanol and biogas from spruce wood by NMMO-pretreatment and rapid fermentation and digestion. *Bioresour. Technol.*, 102, 7879–7886, 2011.

Shafiei, M., Kabir, M.M., Zilouei, H., Sarvari, H.I., Karimi, K., Techno-economical study of biogas production improved by steam explosion pretreatment. *Bioresour. Technol.*, 148, 53–60, 2013.

Shafiei, M., Karimi, K., Zilouei, H., Taherzadeh, M.J., Economic impact of NMMO pretreatment on ethanol and biogas production from pinewood. *Biomed Res. Int.*, 148, 53–60, 2014. Article ID:320254.

Shaheen, R., Shirley, M., Jones, D., Comparative fermentation studies of industrial strains belonging to fourspecies of solvent-producing clostridia. *J. Mol. Microbiol. Biotechnol.*, 2, 1, 115–124, 2000.

Soni, B.K., Das, K., Ghose, T.K., Bioconversion of agro-wastes into acetone butanol. *Biotechnol. Lett*, 4, 19–22, 1982.

Survase, S.A., Van Heiningen, A., Granstrom, T., Continuous bio-catalytic conversion of sugar mixture to acetone-butanol-ethanol by immobilized *Clostridium acetobutylicum* DSM 792. *Appl. Microbiol. Biotechnol.*, 93, 2309–2316, 2012.

Surisetty, V.R., Dalai, A.K., Kozinski, J., Alcohols as alternative fuels: an overview. *Appl. Catal. A Gen.*, 404, 1–11, 2011.

Szulczyk, K.R., Which is a better transportation fuel – butanol or ethanol? *Int. J. Energy Environ.*, 1, 501–12, 2010.

Taherzadeh, M. and Karimi, K., Pretreatment of lignocellulosic wastes to improve ethanol and biogas production: A Review. *Int. J. Mo. Sci.*, 9, 1621–1651, 2008.

Tigunova, E.A., Shul'ga, S.M., Blium Ia, B., Alternative type of fuel–biobutanol. *Tsitol. Genet.*, 47, 51–71, 2013.

Tummala, S.B., Welker, N.E., Papoutsakis, E.T., Design of antisense RNA constructs for downregulation of the acetone formation pathway of *Clostridium acetobutylicum*. *J. Bacteriol.*, 185, 1923–34, 2003.

United States Energy Information Administration (USEIA), *International Energy Outlook 2011*, 2011, http://www.eia.gov/forecasts/ieo/pdf/0484(2011).pdf (accessed January 3, 2012).

Woods, D.R., The genetic engineering of microbial solvent production. *Trends Biotechnol.*, 13, 259–264, 1995.

Wu, Y.D., Xue, C., Chen, L.J., Bai, F.W., Effect of zinc supplementation on acetone -butanol-ethanol fermentation by *Clostridium acetobutylicum*. *J. Biotechnol.*, 165, 18–21, 2013.

Xue, C., Zhao, J.B., Chen, L.J., Bai, F.W., Yang, S.T., Sun, J.X., Integrated butanol recovery for an advanced biofuel: current state and prospects. *Appl. Microbiol. Biotechnol.*, 98, 3463–74, 2014.

Xue, C., Zhao, X.Q., Liu, C.G., Chen, L.J., Bai, F.W., Prospective and development of butanol as an advanced biofuel. *Biotechnol. Adv.*, 31, 1575–1584, 2013.

Xue, C., Zhao, J., Lu, C., Yang, S.T., Bai, F., Tang, I.C., High-titer n-butanol production by *clostridium acetobutylicum* JB200 in fed-batch fermentation with intermittent gas stripping. *Biotechnol. Bioeng.*, 109, 2746–2756, 2012.

Zhang, Y., Hou, T., Li, B., Liu, C., Mu, X., Acetone-butanol-ethanol production from corn stover pretreated by alkaline twin-screw extrusion pretreatment. *Bioprocess. Biosyst. Eng.*, 37, 913–921, 2014.

Zheng, Y.N., Li, L.Z., Xian, M., Ma, Y.J., Yang, J.M., Xu, X., He, D.Z., Problems with the microbial production of butanol. *J. Ind. Microbiol. Biotechnol.*, 36, 1127–1138, 2009.

Zverlov, V.V., Berezina, O., Velikodvorskaya, G.A., Schwartz, W.H., Bacterial acetone and butanol production by industrial fermentation in the Soviet Union: use of hydrolyzed agricultural waste for biorefinery. *Appl. Microbiol. Biotechnol.*, 71, 587–97, 2006.

6

Biomethanol as a Potential Source for Sustainable Energy

Prakash Kumar Sarangi

College of Agriculture, Central Agricultural University, Imphal, Manipur, India

Abstract
Worldwide energy sources are dependent on fossil-based resources, which are in diminishing supply. The production of various chemicals from fossil-based resources increase in our daily life. These products are not only unsustainable. They also create drastic environmental problems. There is global concern about the shifting from conventional, non-renewable sources of energy to sustainable and renewable sources. Biomass acts as crucial feedstock to provide the necessary fuel, energy, and chemicals. Considering global bioeconomy, biomass has evolved as green and clean feedstock for the utilization of bioenergy. Waste biomass acts as a sustainable source to produce bioenergy toward future energy carriers. Biomethanol from biomass, acting as alternative transportation fuel, can be obtained by microbial activities. Methanotrophic bacteria act as a vital agent for the conversion of methane into methanol. This chapter summarizes the scope of microorganisms to produce biomethanol from biomass. Various industrial applications and production pathways of biomethanol are also comprehensively described. Challenges and perspectives in the usage of several technologies in biomethanol production are briefly summarized.

***Keywords*:** Methane, methanotrophic bacteria, biomethanol, biomass, waste

6.1 Introduction

There is a major global concern in the application of green and clean energy sources into automobile and industrial sectors (Okonko *et al.*, 2009; Suntana *et al.*, 2009). Due to the diminishing supply of non-renewable energy and its adverse environmental impacts, alternative and clean energy is needed

Prakash Kumar Sarangi. Waste Biomass to Alternative Fuels, (99–108) © 2025 Scrivener Publishing LLC

for the future. Hence, a sustainable process technology must be developed for the utilization of waste biomass, which would not generate any harmful pollutants for the environment (Sarangi *et al.*, 2020b), being a challenge for researchers around the world (Bakul, 2011; Sarangi, 2019). Thus, waste biomass can be used for the production of valuable chemicals, green energy, and other biobased materials, along with value-added biochemicals toward a sustainable economy and development (Sarangi *et al.*, 2017; Subudhi *et al.*, 2025; Sarangi *et al.*, 2023; Sarangi *et al.*, 2023a; Sarangi *et al.*, 2023b). Hence, green emerging technologies can be emphasized for the conversion of organic wastes into various biofuels for the mitigation of the energy crisis for long-term purposes. Biofuel acts as a cost effective and eco-friendly alternative source for the substitution of fossil fuels. A wide array of by-products from agro-industries such as corn cobs, pineapple waste, sugarcane bagasse, bran, and straw can be utilized as cheap sources for the bioconversion of fuels and value-added biochemicals (Sarangi *et al.*, 2019; Sarangi and Nanda, 2019a; Sarangi and Nanda, 2019b; Sarangi and Nanda, 2019c). A biobased energy system, based on environment-friendly bioprocess technologies, plays a vital role in mitigating fossil-based fuels and in replacing fossil-based fuels and materials. A growing major global concern on energy problems is based on bioenergy utilization, utilizing second-generation biofuels. The establishment of clean, green, sustainable, and benign technologies to produce fuels, power, heat, and value-added chemicals is a global target for academicians and researchers in the bioenergy sector (Sarangi and Nanda, 2018; Sarangi and Nanda, 2020). By utilizing microbial communities, waste biomass like crop waste and agro-industrial waste materials are used to supply energy carriers in the transportation sector and electricity storage. Economically and ecologically, these sources are suitable for the above products. Based on specific conversion methods, various fuel products such as bio-oil, biodiesel, syngas, biohydrogen, bioethanol, biobutanol, biomethanol, and biopropanol can be produced from waste biomass, exploring biobased renewable sources toward the mitigation of fossil-based environmental hazards (Nanda *et al.*, 2020; Pasin *et al.*, 2020; Siang *et al.*, 2020; Bhatia *et al.*, 2020; Sarangi *et al.*, 2020).

Biomethanol is a potential biofuel for future bioeconomy purposes (Table 6.1). Methanol can be focused on its production from waste biomass by utilizing microbial communities. Having various applications, methanol has expanded prominent sectors toward the production of high value chemicals and materials. It can be regarded as a valuable target for the waste-to-chemical process in chemical and fuel industries. This chapter focuses on methanol's applications in various sectors and the biological production technologies of biomethanol from various biomass and microorganisms.

Table 6.1 Common fuel properties of methanol among major bioalcohols (Sarangi and Nanda, 2018; Sarangi *et al.*, 2020).

Characteristics	Methanol	Ethanol	Butanol
Formula	CH_3OH	CH_3CH_2OH	C_4H_9OH
Flash point (°C)	11	13	28. 9
Air/fuel ratio	6.5	9.0	11.2
Auto-ignition temperature (°C)	435	365	343
Higher heating value (kJ/g)	20	29.7	37.3
Viscosity at 25 °C (mPa.s)	0.54	1.07	2.6
Energy density (MJ/kg)	19.6	26.9	33.1
Heat of vaporization (MJ/kg)	1.20	0.92	0.43
Density at 20 °C (g/m^3)	0.797	0.789	0.81
Research octane number	136	129	96

6.2 Application of Methanol

Methanol is regarded as safe fuel due to its low toxicity and high bioconversion rate (Malcolm, 1999). It also has a greater octane number than gasoline (Fatih *et al.*, 2011). Thus, it is considered as a promising alternative for the future fuel crisis.

6.3 Production of Biomethanol

Biomass can be converted into methanol using two important methods, which are thermochemical process and the biochemical process. The former deals with the gasification process, converting waste biomass into synthesis gas (syngas), which is then converted into methanol by a suitable catalyst at a high temperature and pressure. The production of biomethanol from sugar cane bagasse and corn cobs were 5.93% and 0.67%, respectively (Chia, 2011). Also, the production of biomethanol from bio-oil has been detected from CO-rich bio-syngas (Xu *et al.*, 2011). Figure 6.1 shows the production and applications of Biomethanol. Figure 6.2 shows sources for the production of methanol.

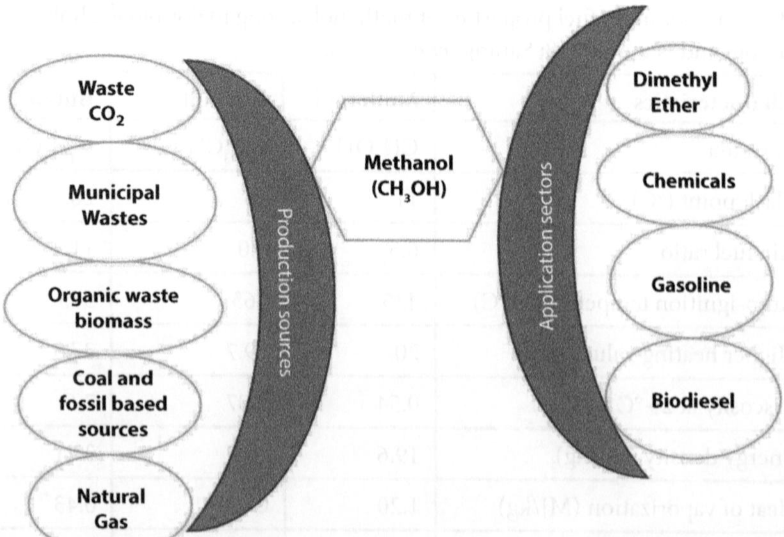

Figure 6.1 The production and applications of biomethanol.

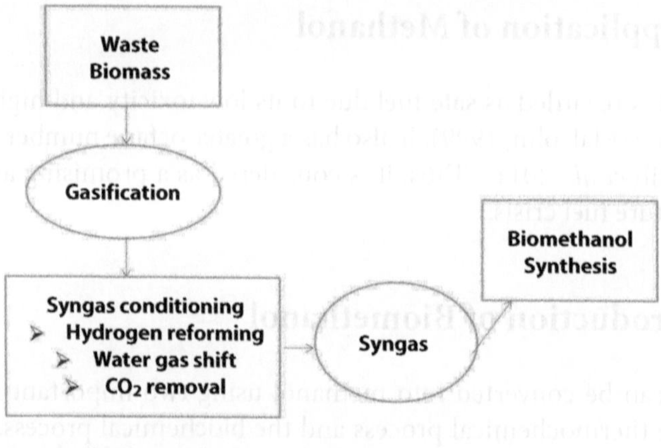

Figure 6.2 Sources for the production of methanol.

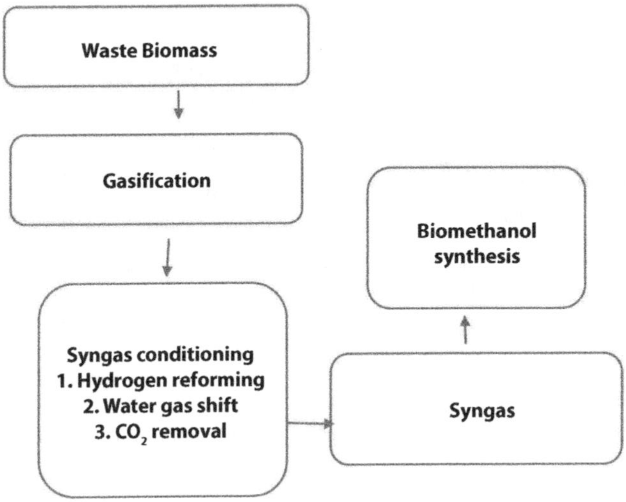

Figure 6.3 An overview of a major biomethanol production process from waste biomass through the thermochemical process.

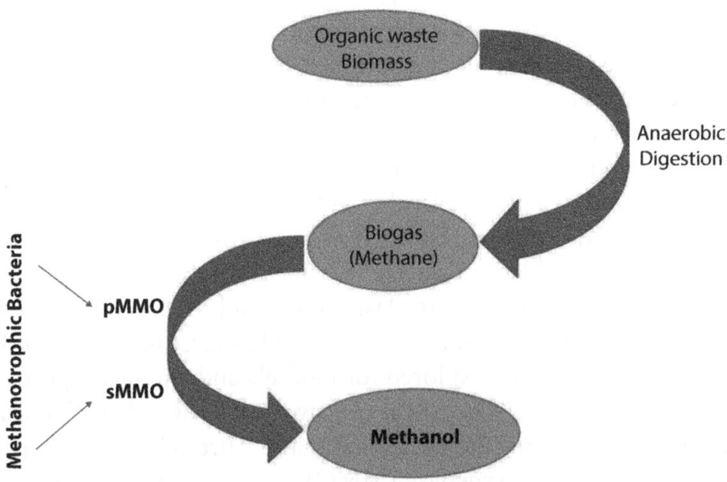

Figure 6.4 A microbial conversion of biomass into biomethanol.

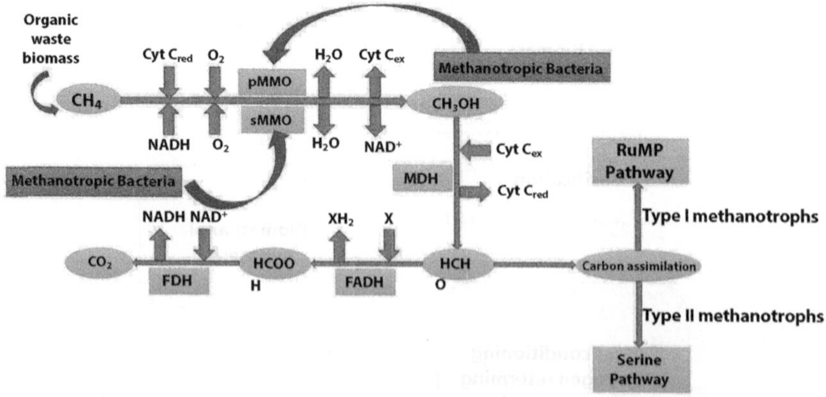

Figure 6.5 Biochemistry for the production of biomethanol through biological methods.

Biochemical methods help the utilization of waste biomass and anaerobic digestion to methane, which is converted into biomethanol by the help of methanotrophic bacteria. An overview of a major biomethanol production process from waste biomass through thermochemical process is shown in Figure 6.3. Agro-residues, crop residues, and by-products such as husks, straw, and bagasse are considered as important sources for biomethanol production. Methanotrophic bacteria help in the conversion of methane as the sole source of carbon and energy to methanol. Figure 6.4 shows the microbial conversion of biomass into biomethanol. Figure 6.5 shows the biochemistry in producing biomethanol through biological methods.

When processing a pineapple, about 60% of a regular pineapple's weight was an unutilized by-product (Singh *et al.*, 2018), but it can be utilized in obtaining value-added products. Microbes have been exploited toward biomethanol production since the last three decades (Hanson and Hanson, 1996). Lignocellulose materials act as a valuable source for hexose and pentose sugars, which can be explored for biofuels and chemicals production (Wassel and Dittmer, 2005). There are two categories of methanotrophs based on the conversion efficiency of methane. Through the catalytic function of methane monooxygenase (MMO) enzymes, such microbes aid in converting methane into methanol through the oxidation process. These microbes can attain maximum recovery of biomethanol through more extensive research for biomethanol production.

6.4 Conclusion

As far as energy security and environmental sustainability are concerned, waste biomass can act as a potential alternative to biofuel and biochemical production. Microbial biomass with waste biomass can support a sustainable source for future biofuel production. Methanol has many applications in energy, fuel, and chemical sectors. Biomethanol from biobased sources has great potential in automobile and industrial sectors, supplying fuel and energy for the future. There are various production technologies, such as thermochemical and biological conversion processes that are employed for methanol production depending on the feedstock. A wide array of biochemicals can be processed in secondary and tertiary manners from methanol, sharing major proportions as bioenergy and fuel carriers. Methanotrophic bacteria are used biologically to produce methanol from methane, which is ultimately obtained from waste biomass through the anerobic digestion process. Waste biomass is regarded as an alternative source to produce biomethanol with the aid of methanotrophic bacteria via methane. The bioprocessing of waste biomass through microorganisms toward biomethanol has many challenges for large-scale recovery.

Process parameters such as the mode of operation, biomass properties, and process conditions affect the production rate of biomethanol. Parameters like temperature, heating rate, conversion processes, particle size, and residence time also have great roles in final product recovery. The economics of biomethanol not only depends on production parameters and conditions but also on the purity of the final product. Novel strain and standardization of cultural conditions may fulfil the problems in the final recovery of biomethanol. Thus, process developments for the large-scale production of methanol are the focus for future applications in fuel and chemical sectors, using biomethanol as a substantial energy carrier, mitigating the environmental crisis due to fossil-based sources.

References

Bakul, C.D., *Prospects for methanol production*, ASM Press, Washington, 2011.

Chia, S.K., *Production of bio-methanol from agricultural waste by pyrolysis method*, Department of Chemical and Process Engineering, National University of Malaysia, 2011, Malaysia. (Thesis).

Fatih, D.M., Balat, M., Balat, H., Biowastes to biofuels. *Energy Convers. Manage.*, 52, 1815–1828, 2011.

Hanson, R.S. and Hanson, T.E., Methanotrophic bacteria. *Microb. Reviews*, 60, 439–471, 1996.

Malcolm, P.I., Evaluation of the fate and transport of methanol in the environment. *Am. Methanol. Inst.*, Report No.: 3522-002, 69, 1999.

Okonko, I.O., Egwame, R.A., Fajobi, E.A., Nkang, A.O., Iheakanwa, C.I., Ogunnusi, T.A., *et al.*, Current trends in bio-fuel production and its use as an alternative energy security. *Environ. Agric. Food Chemistry*, 8, 1233–60, 2009.

Sarangi, P.K., Nanda, S., Vo, D.V.N., Recent Advances in Steam Reforming of Glycerol for Syngas Production, in: *Biorefinery of Alternative resources: Targeting green fuels and platform chemicals*, S. Nanda, D.V.N. Vo, P.K. Sarangi (Eds.), pp. 399–426, Springer Nature, 2020, ISBN: 978-9811518034.

Sarangi, P.K., Vivekanand, V., Mohanakrishna, G., Pattnaik, B., Muddapur, U., Aminabhavi, T.M., Biorefinery of agricultural byproducts for the production of phenolic compounds: A bioeconomic perspective. *J. Clean. Prod.*, 414, 137460, 2023b, https://doi.org/10.1016/j.jclepro.2023.137460.

Sarangi, P.K. and Nanda, S., Recent developments and challenges of acetone-butanol-ethanol fermentation, in: *Recent Advancements in Biofuels and Bioenergy Utilization*, P.K. Sarangi, S. Nanda, P. Mohanty (Eds.), pp. 111–123, Springer Nature, Singapore, 2018, ISBN No.: 978-981-13-1306-6.

Sarangi, P.K. and Nanda, S., Recent advances in consolidated bioprocessing for microbe-assisted biofuel production, in: *Fuel Processing and Energy Utilization*, S. Nanda, P.K. Sarangi, D.V.N. Vo (Eds.), pp. 141–157, CRC Press, Florida, USA, 2019, ISBN No.: 978-1-138-59320-6.

Sarangi, P.K. and Nanda, S., Bioconversion of agro-wastes into phenolic flavour compounds, in: *Biotechnology for Sustainable Energy and Products*, P.K. Sarangi and S. Nanda (Eds.), pp. 266–284, IK International Publishing House, India, 2019a.

Sarangi, P.K. and Nanda, S., Valorization of pineapple wastes for biomethane generation, in: *Biogas Technology*, S. Mishra, T.K. Adhya, S.K. Ojha (Eds.), pp. 169–180, New India Publishing Agency, New Delhi, India, 2019b, ISBN:978-93-87973-52-7.

Sarangi, P.K., Sustainable Utilization of Agro Wastes using Microbial Resources: A Future Perspective, in: *Microbe Assisted Sustainable Agriculture*, P.K. Sarangi and B. Sinha (Eds.), pp. 257–268, Overseas Press India Private Limited New Delhi, India, 2019, ISBN: 978-81-939968-9-8.

Sarangi, P.K. and Nanda, S., Biohydrogen Production Through Dark Fermentation. *Chem. Eng. Technol.*, 43, 4, 601–612, 2020.

Sarangi, P.K., Nanda, S., Vo, D.V.N., Technological Advancements in the Production and Application of Biomethanol, in: *Biorefinery of Alternative resources: Targeting green fuels and platform chemicals*, S. Nanda, D.V.N. Vo, P.K. Sarangi (Eds.), pp. 127–140, Springer Nature, Singapore, 2020, ISBN: 978-9811518034.

Sarangi, P.K., Singh, T.A., Singh, N.J., Pineapple as potential crop resource: Perspective and value addition, in: *Food Bioresources and Ethnic Foods of Manipur*, pp. 83–91, Empyreal Publishing House, Northeast, India, 2019.

Sarangi, P.K., Srivastava, R.K., Singh, A.K., Sahoo, U.K., Prus, P., Sass, R., Municipal-Based Biowaste Conversion for Developing and Promoting Renewable Energy in Smart Cities. *Sustainability*, 15, 12737, 2023. https://doi.org/10.3390/su151712737.

Sarangi, P.K., Subudhi, S., Bhatia, L., *et al.*, Utilization of agricultural waste biomass and recycling toward circular bioeconomy. *Environ. Sci. Pollut. Res.*, 30, 8526–8539, 2023a. https://doi.org/10.1007/s11356-022-20669-1.

Subudhi, S., Saha, K., Mudgil, D., Srivastava, R.K., Sarangi, P.K., Biomethanol production from renewable resources: a sustainable approach. *Environ. Sci. Pollut. Res.*, 32, 7432–7448, 2025. https://doi.org/10.1007/s11356-023-29616-0.

Suntana, A.S., Vogt, K.A., Turnblom, E.C., Ravi, U., Bio-methanol potential in Indonesia: forest biomass as a source of bio-energy that reduces carbon emissions. *Appl. Energy*, 86, 215–21, 2009.

Wassel, C.S. and Dittmer, T.P., Are subsidies for biodiesel economically efficient? *Energy Policy*, 34, 3993–4001, 2005.

Xu, Y., Ye, T., Qiu, S., Ning, S., Gong, F., Liu, Y., *et al.*, High efficient conversion of CO2-rich bio-syngas to CO-rich bio-syngas using biomass char: a useful approach for production of bio-methanol from bio-oil. *Bioresour. Technol.*, 102, 6239–45, 2011.

About the Author

Dr. Prakash Kumar Sarangi

Dr. Prakash Kumar Sarangi is a scientist at Central Agricultural University in Imphal, Manipur, India. He has a Ph.D. in Microbial Biotechnology from at Ravenshaw University, Cuttack, India; M.Tech. degree from the Indian Institute of Technology Kharagpur, India. Dr. Sarangi's current research is focused on bioprocess engineering, biofuels, biochemicals, biomaterials, fermentation technology, and post-harvest engineering and technology. He has more than 13 years of teaching and research experience in biochemical engineering, microbial biotechnology, downstream processing, food microbiology and molecular biology. He has served as a reviewer for many international journals and has authored more than 130 peer-reviewed research articles and 75 book chapters. Dr. Sarangi has edited many books from different publishers like Springer Nature, CRC Press and Elsevier. Dr. Sarangi serves as editorial board members and guest editors in reputed journals. He is associated with many scientific societies as a fellow member *(Society for Applied Biotechnology)* and life member *(Biotech Research Society of India; Society for Biotechnologists of India; Association of Microbiologists of India; Orissa Botanical Society; Medicinal and Aromatic Plants Association of India; Indian Science Congress Association; Forum of Scientists, Engineers & Technologists; International Association of Academicians and Researchers; Hong Kong Chemical, Biological & Environmental Engineering Society; International Association of Engineers; and Science and Engineering Institute.)*

Index

Also of Interest

By the same author:

Biorefinery Production of Fuels and Platform Chemicals, Edited by Prakash Kumar Sarangi, ISBN: 9781119724728. A complete biorefinery guide to waste biomass conversion technologies for the production of energy and platform chemicals.

"Advances in Biofeedstocks and Biofuels" series:

Volume 1:
BIOFEEDSTOCKS AND THEIR PROCESSING, Edited by Lalit Kumar Singh and Guarav Chaudhary, ISBN: 9781119117254. The most comprehensive and up-to-date treatment of all the possible aspects for biofeedstock processing and the production of energy from biofeedstocks.

Volume 2:
PRODUCTION TECHNOLOGIES FOR BIOFUELS, Edited by Lalit Kumar Singh and Gaurav Chaudhary, ISBN: 9781119117520. This second volume in the "Advances in Biofeedstocks and Biofuels" series focuses on the latest and most up-to-date technologies and processes involved in the production of biofuels.

Volume 3:
LIQUID BIOFUEL PRODUCTION, Edited by Lalit K. Singh and Gaurav Chaudhary, ISBN: 9781119459873. Focusing on renewable and carbon-cutting alternative fuel sources. this third volume in the "Advances in Biofeedstocks and Biofuels" series focuses on the latest and most up-to-date technologies and processes involved in the production of liquid biofuels.

Volume 4:
PRODUCTION TECHNOLOGIES FOR SOLID AND GASEOUS BIOFUELS, Edited by Lalit Kumar Singh and Gaurav Chaudhary, ISBN: 9781119785828. This latest volume in the series, "Advances in Biofeedstocks and Biofuels," offers the most up-to-date and comprehensive coverage available for the production technologies for solid and gaseous biofuels.

Other Related Titles

BIOFUEL EXTRACTION TECHNIQUES, Edited by Lalit Prasad, Subhalaxmi Pradhan, and S.N. Naik, ISBN: 9781119829324. Written and edited by a team of world-class engineers, this groundbreaking new volume presents the current state of the art in the processing of feedstocks for biofuel production and end use application, extraction techniques, valorization of biofuel waste, and recent advances in biorefineries and the recycling industries.

LIQUID BIOFUELS: Fundamentals, Characterization, and Applications, Edited by Krushna Prasad Shadangi, ISBN: 9781119791980. From the selection and pretreatment of raw materials to design of reactors, methods of conversion, selection of process parameters, optimization, and production of various types of biofuels to the industrial applications for the technology, this is the most up-to-date and comprehensive coverage of liquid biofuels for engineers and students.

BIODIESEL TECHNOLOGY AND APPLICATIONS, Edited by Inamuddin, Mohd Imran Ahamed, Rajender Boddula, and Mashallah Rezakazemi, ISBN: 9781119724643. This outstanding new volume provides a comprehensive overview on biodiesel technologies, covering a broad range of topics and practical applications, edited by one of the most well-respected and prolific engineers in the world and his team.

BIOFUEL CELLS: Materials and Challenges, Edited by Inamuddin, Mohd Imran Ahamed, Rajender Boddula, and Mashallah Rezakazemi, ISBN: 9781119724698. This book covers the most recent developments and offers a detailed overview of fundamentals, principles, mechanisms, properties, optimizing parameters, analytical characterization tools, various types of biofuel cells, edited by one of the most well-respected and prolific engineers in the world and his team.

Printed and bound by CPI Group (UK) Ltd, Croydon, CR0 4YY
17/09/2026
14964245-0008